# CONTENTS

D0375659

# GETTING STARTED

Moving from an idea to a physical object is the promise of 3D printing.

If you want to design three-dimensional objects on the computer, then do it. With access to a 3D printer and some free design software, your first project will take less than one minute.

Look around. Most of the objects you see started as a box or a can, often with holes drilled in them or pieces sticking out. You can make things like that.

Do you have an idea for a new invention? You can take the next step in realizing that potential. Using an apprentice approach—do the work and learning will occur—you will go through every step in the design of 15 toys. Whether your goal is building toys, making stuff, or learning about 3D printing, going through the projects will show you how to create and manipulate 3D objects.

Consumer-grade (read: affordable) 3D printers create objects by extruding layers of hot plastic, sort of like a computer-controlled hot glue gun. The most commonly used plastics are ABS (the same material as in Legos) and PLA (a biodegradable alternative). PLA printers tend to be easier to use, since they don't require heated beds or temperature-controlled chambers.

Expensive commercial printers can produce objects using metal, food, or biological material. The expiration of a key laser sintering patent in 2014 (US patent number 5,597,589) opens the door to improvements in speed, cost, quality, and material capabilities of 3D printers.

Printing takes time. A decorative ring for your finger will require 10 to 15 minutes to print. Similar to paper printers, a 3D printer has different quality levels for the printed product. A less detailed object prints faster. Depending on the printer settings, the inside areas of a printed object can be hollow (0 percent fill), solid (100 percent fill), or anywhere between. The printer's default settings are normally adequate for the projects in this book.

If you are not ready to purchase a printer, there are several ways to produce a real-world object from a computer-generated file for a small fee. Some libraries have 3D printers available for their patrons. Autodesk 123D Design has a **Send to 3D Print Web Service** button built into the software. Commercial 3D printing services (search the web for "3D printing service") can even produce your items in silver. To find a local printer, visit www.3dhubs.com, which lists thousands of users worldwide who are willing to print projects.

## Downloading the Software

Before you begin, you will need to download and install free design software for Windows or Mac: either 123D Design (www.123dapp.com/design) or SketchUp (www.sketchup.com). *The more detailed instructions in this book use 123D Design.* Follow the instructions and install the software on your computer.

Each program is available for both Windows and Mac computers. The 123D Design instructions were prepared on a Windows laptop computer (the lowest-cost laptop from a big box store) and the SketchUp instructions were prepared on a Mac laptop, but they should work with either operating system. Small differences between the Windows and Mac versions—or future updates to the software—may change the program's configuration slightly. If this happens, look for an option with the same or a similar name.

For SketchUp, you will need to do two additional setup tasks. When installed, open the program. Select **Choose Template** and select **Woodworking—Millimeters** then left-click **Start Using SketchUp** (at the lower right corner of the box with the template selections).

One time only, you will need to do this step. Go to the top of the screen, find the second icon from the right (**SketchUp Extension**), and left-click with your mouse. In the white box in the upper right corner, type STL. Left-click on the red magnifying glass to the right of the white box. From the results, left-click **SketchUp STL**. On the next screen, choose **Install**. Next, choose **Accept**. Click on **Install** and you are ready.

Files for all of the projects in this book are available at no cost through the website www.MisterEngineer.com. More than 100,000 designs for printing are available (at no charge) through a website for sharing things, www.Thingiverse.com.

# 1

# Domino-Style Blocks

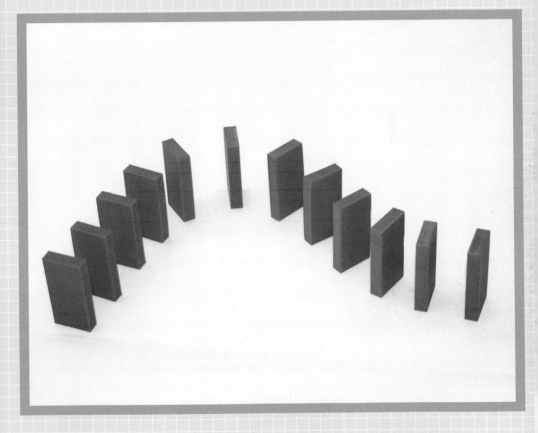

This is an introductory project so you can get acquainted with the basics of 3D design. To create domino-style blocks using 123D Design, you will select a box and describe the dimensions of a domino-shaped block.

Go to **Primitives** and drop down to **Box**.

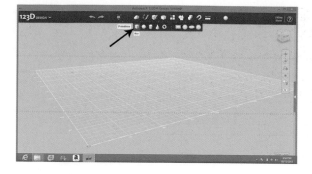

Left-click and release. A cube will appear. Move the cursor to the lower left area of the screen.

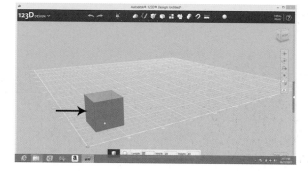

Type 8, then press the **Tab** key. Type 50, then press the **Tab** key. Type 25. Press **Enter**. This will create a box that is 8 mm by 50 mm by 25 mm.

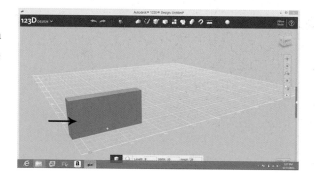

Move the cursor to **123D** in the upper left corner. On the drop-down menu, find **Export STL** and you are ready to print your first project.

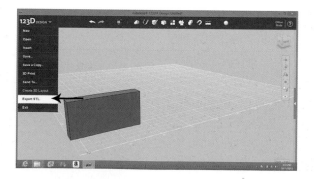

Name the file "Block" and the ".stl" will be attached. Now you can send "Block.stl" to your 3D printer and a domino block will be created. This is called a "print" file and it is written in a language 3D printers can understand.

To save your design file (in case you want to add other details later), move the cursor to **123D** and drop down to **Save**.

Print the block.

## MAKING DOMINO-STYLE BLOCKS USING SKETCHUP

This introductory project can also be created using SketchUp. Start with a new file. Go to **Camera**, then drop down to **Standard Views**, then **Front**. Left-click and release.

Choose the **Shape** icon. Left-click and release. Drop down to **Rectangle**. Left-click and release.

Move the pointer to the intersection of red and blue lines. Left-click and hold. Move the cursor right and slightly up. Type 50,25 and press **Enter**. Release the left-click.

Choose the **Zoom** icon. Left-click and release. Move the pointer to the rectangle. Scroll until the rectangle fills most of the work area.

Move the cursor to the **Push/Pull** icon. Left-click and release.

Move the cursor inside the rectangle, close to the bottom. Left-click and hold. Move the cursor up slightly. Type 8. Press **Enter**, then release the left-click.

Go to the **Select** icon (first on left), left-click and release. Go to **Edit**, then drop down to **Select All**. Left-click and release.

Go to the menu at the top and choose **File**, then drop down to **Export STL**. Left-click. This file can be used to print a domino block.

To save the design file, choose **File**, then **Save**.

# 2

# Spinning Button

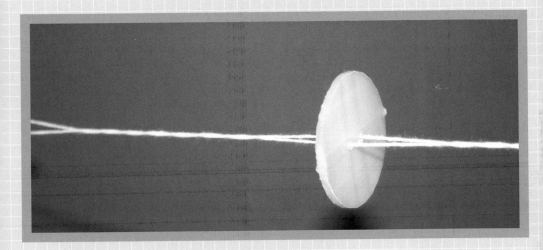

To create a spinning button, you will draw a circle, make that circle thick, then cut two holes in the thick circle.

Start by moving the cursor to the **Sketch** icon.

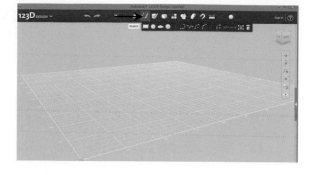

In the drop-down menu, move the cursor to **Circle**.

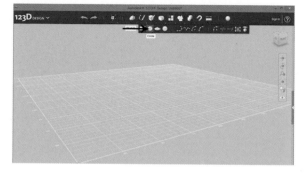

Left-click, and move the cursor down onto the grid.

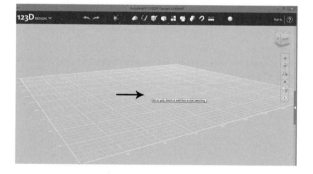

Left-click two times. The first click will result in a check mark. The second click will produce a small dot.

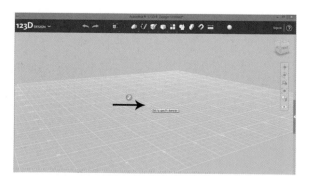

Move the cursor. A box will appear with the message "Click to specify diameter."

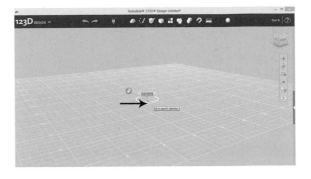

Type 50 and "50" will appear in the box.

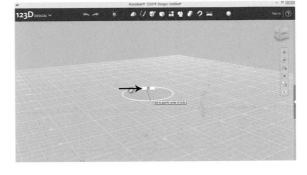

Left-click. Move the cursor to the **Construct** icon at the top of the page.

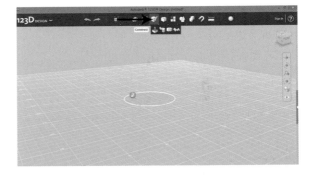

Drop down to the **Extrude** option. It is the far-left icon on the drop-down menu.

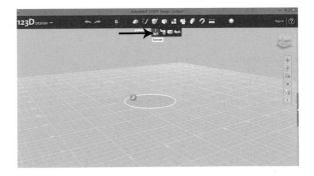

Left-click on the **Extrude** option and the check mark will go away.

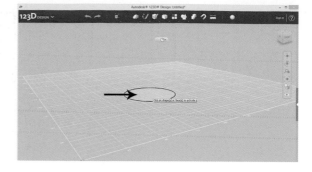

Move the cursor inside the circle.

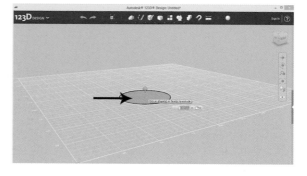

Left-click inside the circle. A box with "0.00 mm" will appear. Type 3 and "3" will appear in the box. Left-click.

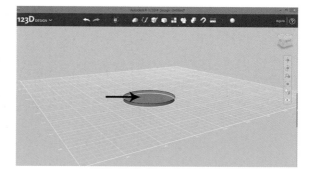

Move the cursor to the upper right corner of the screen. Place the cursor on top of the cube. This cube indicates the view direction.

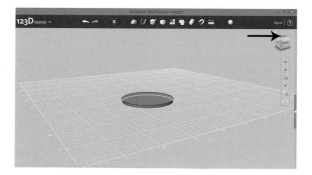

Left-click and you will now view the button from the top.

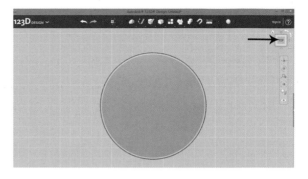

Move the cursor to the **Sketch** icon at the top of the screen.

Drop the cursor down to the **Circle** icon.

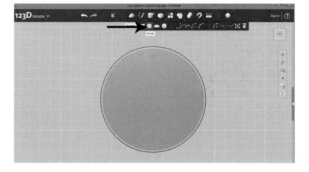

Left-click. Move the cursor down into the circle. There is a tiny dot in the center of the circle. Move the cursor right from the center to the first intersection of grid lines.

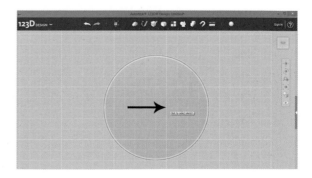

Left-click two times, then move the cursor. A box with a number will appear.

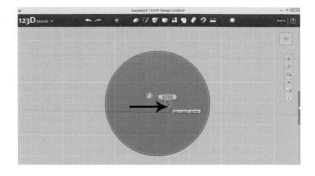

Type 3 and it will appear in the box. This is the size of one of the holes in your button.

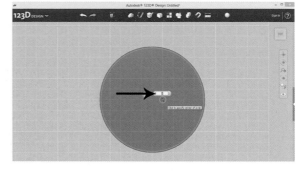

Left-click. Move the cursor to the **Construct** icon at the top of the page.

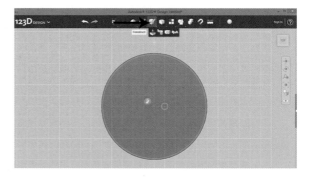

Drop down to the **Extrude** icon.

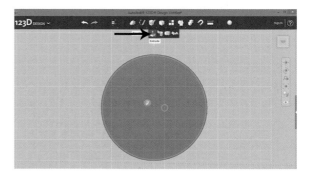

Left-click.

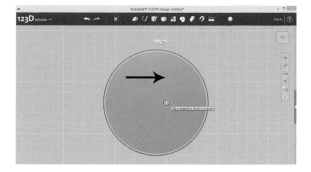

Move the cursor inside the small circle and left-click.

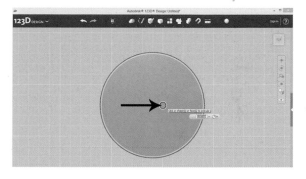

Type −3 and left-click. The "−" sign makes a hole "into" the button. If you had typed −2 and clicked, it would have produced an indentation, not a hole through the button. If you had typed 3 and clicked, a small cylinder would have protruded outward from the button.

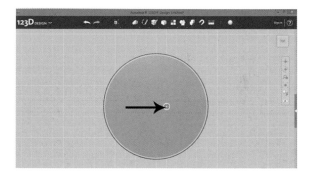

Move the cursor to the **Sketch** icon at the top.

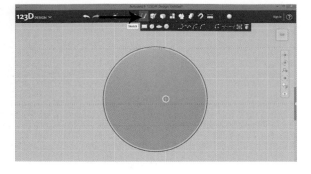

Drop the cursor down to the **Circle** icon.

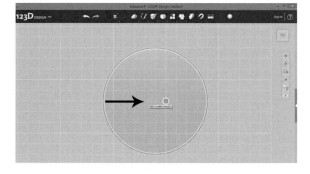

Left-click. Move the cursor into the large circle, left of the center dot at the first intersection of crossing grid lines.

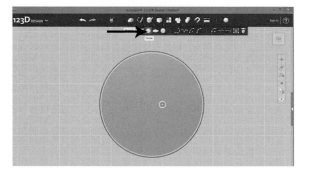

Left-click two times, then move the cursor slightly.

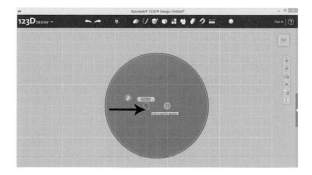

Type 3 and note that "3" appears in the box. Left-click.

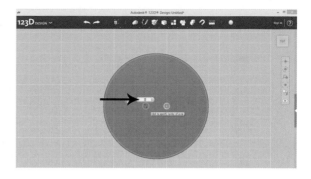

Move the cursor to the **Construct** icon.

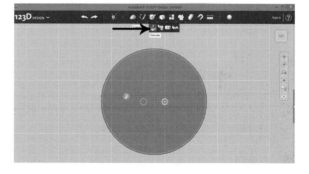

Drop the cursor down to the **Extrude** icon.

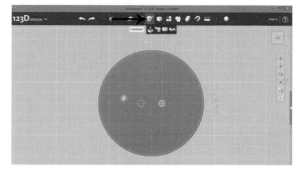

Left-click, then move the cursor into the large circle.

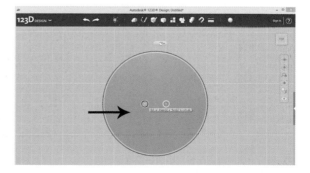

Move the cursor into the small left circle and left-click. Type −3 and "-3" will appear in the box.

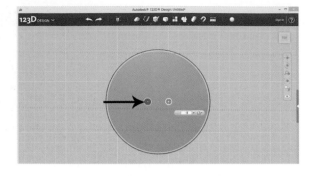

Left-click.

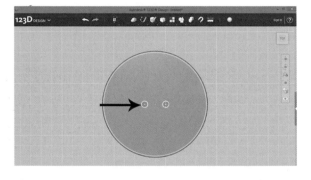

Move the cursor toward the cube in the upper right corner. As the cursor approaches the cube, a small house will appear on the upper left corner of the cube. Left-click on the little house.

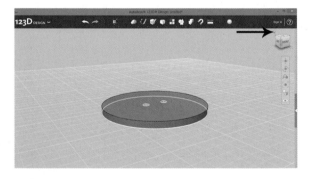

Move the cursor to the upper left, over the **123D** in the corner.

Move the cursor down to **Export STL** and save your file for printing.

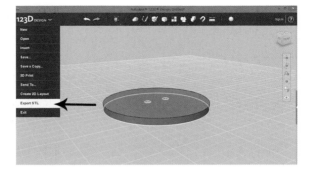

To save the design file, move the cursor to **123D** in the upper left corner and drop down to **Save**.

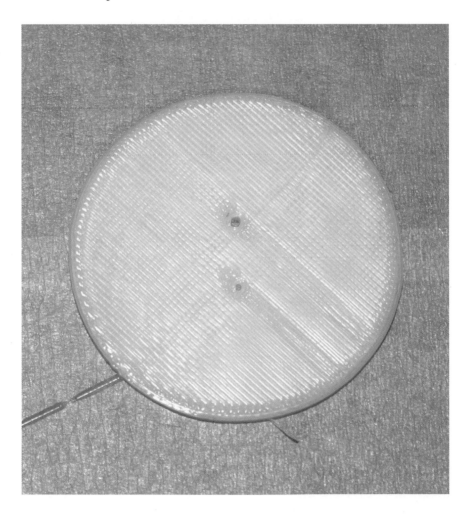

Obtain a piece of string about 1 meter (3 feet) long. Thread the string through the button.

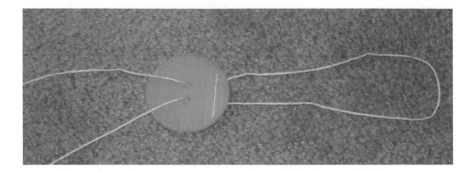

Tie the ends of the string together. Now, holding the ends of the string, spin the button about 20 revolutions. Gently pull and relax the ends of the string, and the button will continue to spin.

## MAKING A BUTTON USING SKETCHUP

To create your button, go to the fifth icon from the left at the top of the screen (**Shape**). Left-click and release. Drop down to **Circle**. Left-click, release, and drag the circle to the intersection of red, green, and blue lines. Left-click and hold. Move the cursor slightly. Notice the box in the lower right that says "Radius" and displays a number. Type 25 and depress the **Return** key. Release the left-click. Move the cursor to the magnifying glass at the top of the screen (sixth from right). Left-click and release. Place the magnifying glass over the circle. Scroll until the circle gets larger.

Move the cursor to the top of the screen, sixth from the left (**Push/ Pull**), then left-click and release. Move the cursor inside the circle (circle will fill with tiny black dots). Left-click and hold while moving the mouse downward. When the box in the lower right ("Distance") is "3.0mm," release the left-click. You have now created a button and given it thickness.

Now make the holes. At the top of the screen, select the **Tape Measure** (near the center of the icons) by left-clicking. Move the cursor toward the intersection of the red, blue, and green lines. When the little dot is at the intersection of the red, green, and blue lines, left-click. Move the dot along the red line (to the right) until the length (box in the lower right) indicates "5.0mm," then left-click. Notice that a small black dot has been placed on the red line. Move your cursor to the **Shape/Circle** icon at the top of the screen, then left-click and release. Move the center of the circle until it "snaps" onto the dot you just made on the red line. Left-click, release and move the cursor until the radius (lower right box) indicates "1.5mm." Left-click. Go to the top of the screen and find the **Push/Pull** icon (sixth from left) then left-click and release. Move the cursor until the small circle is filled with black dots. Left-click, hold and pull the cursor down. When the distance (lower right-hand box) indicates "-3.0mm" release the left-click.

Create another hole. Get the **Tape Measure** and mark a spot 5 mm to the left of the center of the circle. Using the **Shape/Circle** tool, center the circle on your mark and create a 1.5 mm–radius circle. Using the **Push/ Pull**" tool, go downward 3 mm, making another hole.

Go to the icon menu, first icon on the left (an arrow), and left-click. Go to the word menu at the top, find **Edit**, and drop down to **Select All**. Left-click. All surfaces of the button should be covered in tiny black dots.

The button is complete. Go to the menu at the top and choose **File**. From the drop-down menu, choose **Export STL**. The "STL" file you created can be printed on your 3D printer.

To save your design file, choose **File**, then **Save**.

# 3

# Ring

To make a ring, you will draw a circle, make it thick, and cut out a hole for the finger. To make the ring more elaborate, you will draw a decorative shape and make the shape as thick as the ring.

Start by moving the cursor to **Primitives**, then drop down to the **Cylinder** icon.

Left-click, release and drag the cylinder down to the grid.

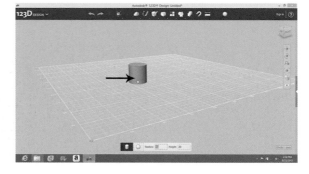

At the bottom of the screen is a box with a highlighted number in the "Radius" area. Type 11 and depress the **Tab** key. The "Height" area is now highlighted. Type 5 and depress the **Enter** key.

Note that the box at the bottom has disappeared.

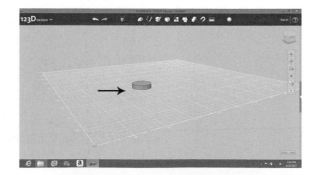

Move the cursor to upper right corner of the screen, pointing to the top of the cube. (The top should be highlighted.)

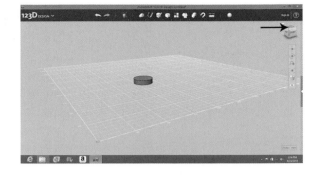

Left-click and release. Move the cursor to the **Sketch** icon and go to the **Spline** icon in the drop-down choices.

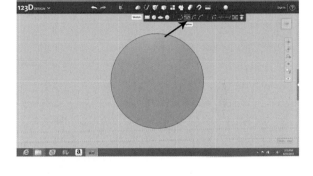

Left-click, release, then move the cursor slightly inside the circle. (The interior of the circle will become almost transparent.)

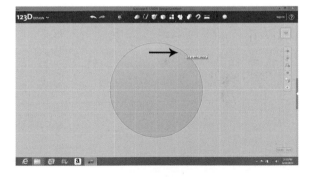

Left-click two times. This will establish the first point in the freehand curve you are drawing.

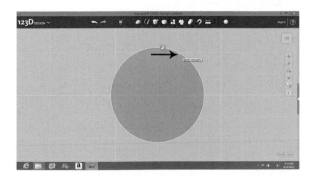

Move outside the circle and left-click once.

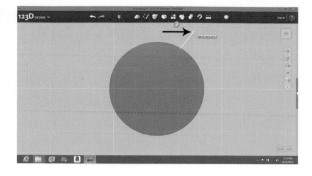

Move to the right and—when you like the position—left-click one time.

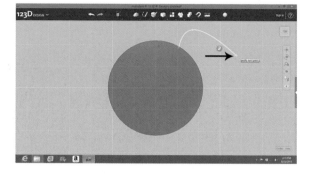

Bring the line inside the circle and left-click once.

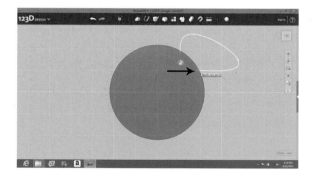

Bring the cursor back to the starting point and left-click one time. The object you have drawn will turn light gray.

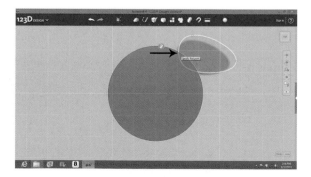

Move the cursor to the **Construct** icon and drop down to the **Extrude** icon.

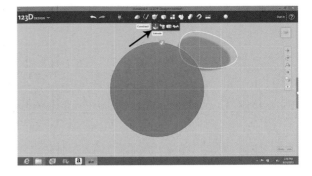

Left-click, release, and move the cursor inside the object you just drew.

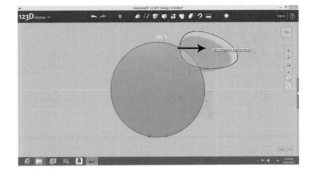

Left-click and release. A box will appear with "0.00 mm" highlighted. Type −5 and move the cursor to the right of the "-5."

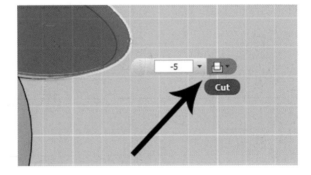

On the drop-down menu, left-click **Merge**. You are adding depth here, not cutting it away.

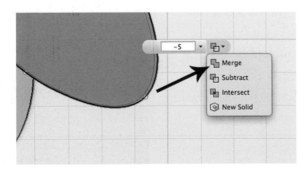

Depress **Enter**. This makes your created object 5 mm thick to match the cylinder, which already has 5 mm of thickness.

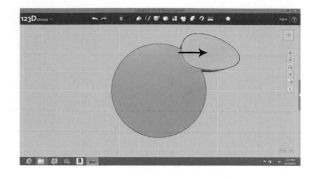

Move the cursor to the **Sketch** icon and drop down to the **Circle** icon.

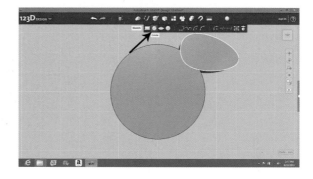

Left-click, release and move the cursor to the center of the circle. Left-click two times, then move the cursor slightly.

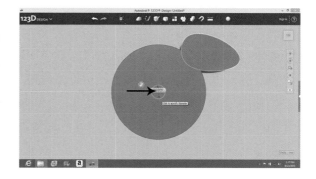

Type 18 and it will appear in the high-lighted box (and the circle will be drawn to a diameter of 18 mm).

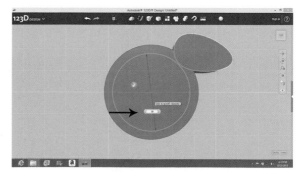

Left-click and release. The message will change from "specify diameter" to "specify center of circle."

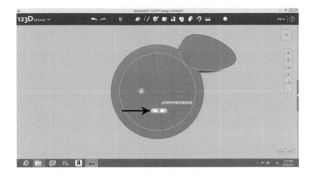

Move the cursor to the **Construct** icon and drop down to the **Extrude** icon.

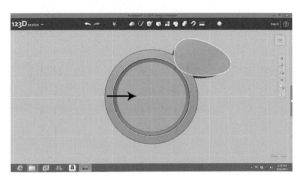

Left-click, release, and move the cursor inside the circle you just drew. Left-click.

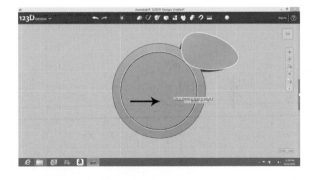

Type −5. (It will appear in the box near the cursor.) Depress the **Enter** key.

Move the cursor to the **123D** in the upper left corner and drop down to the **Export STL** line. Left-click and save your file for printing.

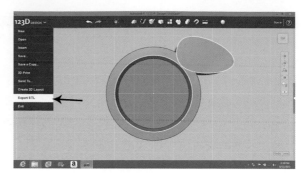

The ring is now ready to print. Use the **Enlarge/Reduce** function of your printer to adjust the fit for different finger sizes.

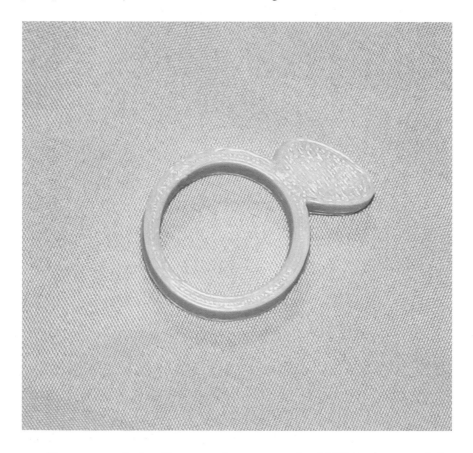

To save your design file, move the cursor to the **123D** in the upper left corner and drop down to **Save**.

# MAKING A RING USING SKETCHUP

To create a ring, go to the fifth icon from the left at the top of the screen (**Shape**). Left-click and release. Drop down to **Circle**. Left-click, release, and drag the circle to the intersection of red, green and blue lines. Left-click and hold. Move the cursor slightly. Notice the box in the lower right that says "Radius" and displays a number. Type 11 and depress the **Return** key. Release the left-click. Move the cursor to the magnifying glass at the top of the screen (sixth from right). Left-click and release. Place the magnifying glass over the circle. Scroll until the circle gets larger.

Move the cursor to the top of the screen, sixth from the left (**Push/Pull**), then left-click and release. Move the cursor inside the circle (circle will fill with tiny black dots). Left-click and hold while moving the mouse downward. When the box in the lower right ("Distance") is "5.0mm," release the left-click.

Go to the fifth icon from the left at the top of the screen (**Shape/Circle**). (**Shape** can be a rectangle, circle, or polygon. When the instructions indicate **Shape/Circle**, drop down to **Circle** if it is currently set to **Rectangle**. When instructions indicate **Shape/Rectangle**, drop down to **Rectangle** if it is set to **Circle**.) Left-click, release and drag the circle to the intersection of red, green, and blue lines. Left-click and hold. Move the cursor slightly. Notice the box in the lower right that says "Radius" and displays a number. Type 9 and depress the **Return** key. Release the left-click.

Move the cursor to the top of the screen, sixth icon from the left (**Push/Pull**) then left-click and release. Move the cursor inside the circle. (The circle will fill with tiny black dots.) Left-click and hold while moving the mouse downward. When the box in the lower right ("Distance") is "-5.0mm," release the left-click. You have created a ring. Next you will add a creative shape to the ring.

Take the cursor to the third icon from the left (**Lines**) and left-click on **Freehand** from the drop-down menu. Move the cursor to an edge on the top of the ring. Left-click, hold, and draw a shape of your choice. Bring the line back until it touches the ring, then release the left-click. The area inside the line you just completed should turn white.

Move the cursor to the top of the screen, sixth from the left (**Push/Pull**), then left-click and release. Move the cursor inside the shape you just drew. (The shape will fill with tiny black dots.) Left-click and hold while moving the mouse downward. When the box in the lower right ("Distance") is "5.0mm," release the left-click.

Go to the icon menu, first on the left (an arrow) and left-click. Go to the word menu at the top, find **Edit**, and drop down to **Select All**. Left-click. All surfaces of the ring should be covered in tiny black dots.

Go to the menu at the top and choose **File**, then drop down to **Export STL**. Left-click. This file can be used to print the ring you just designed.

To save the design file, choose **File**, then **Save**.

# 4

# Boxes and Lids

You will design a pirate's chest and an oval box in this chapter. You will take the box shape, make it thick, then cut out a hole. The lid will be printed in the same shape, with a lip to fit inside the hole.

## The Pirate's Chest

Start with the pirate's chest. Go to **Primitives** and drop down to **Box**. Left-click and release.

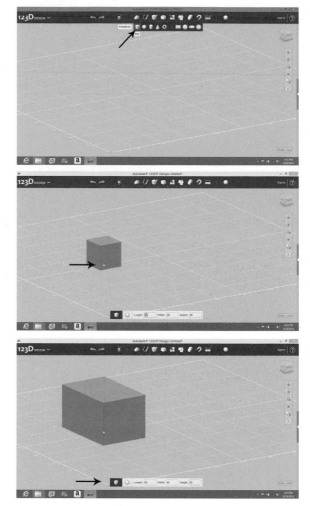

Pull the cube down onto the grid (no mouse clicking).

Type 60 and it will appear in the "Length" box at the bottom of the screen. Press the **Tab** key. Type 45 and it will appear in the "Width" box. Press the **Tab** key. Type 35 and it will appear in the "Height" box.

Press the **Enter** key, then move the cursor to the top of the little cube in the upper right corner.

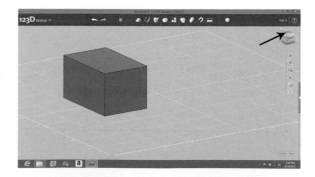

Left-click and release for a top view.

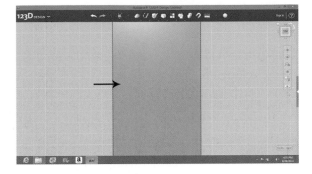

Move the cursor to **Sketch** and drop down to **Rectangle**. Left-click and release.

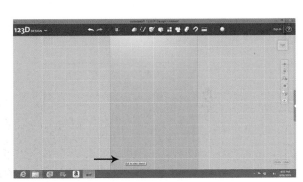

Pull the cursor into the highlighted rectangle. Place the point of the cursor 5 mm right and 5 mm up from the corner of the highlighted rectangle.

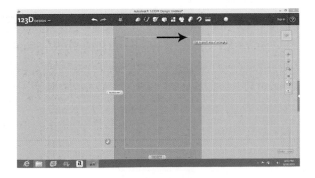

Left-click twice, then move the cursor toward the upper right corner. The lower box will be highlighted. Type 35 and press the **Tab** key. Type 50 and it will appear in the upper box.

Left-click, release, and move the cursor to **Construct**, then drop down to **Extrude**.

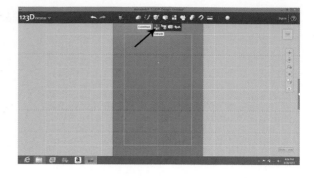

Left-click, then move the cursor inside the rectangle you just created. Left-click again.

Type -30 and it will appear in the box. Move the cursor to the right and you will see the word "subtract" drop down. Usually, the software knows what you want to do. "Subtract" means "to make

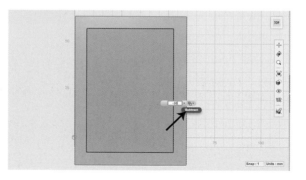

a hole," while "Merge" means "to add something outward." It's a good idea to check this before you accept the operation. (If you wanted the rectangle to extend outward, then you would type 30—no minus sign—and the little symbol to the right of the box would say "Merge.")

Depress the **Enter** key, then move to the **123D** in the upper left corner and drop down to **Export STL**. Left-click and save your file for printing. To save your design file, go to the **123D** and drop down to **Save**, then follow the instructions.

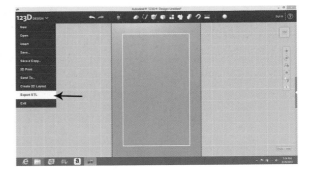

## The Pirate's Chest Lid

Now you'll design a lid for the pirate's chest. Start a new project and move the cursor to the top of the cube in the upper right corner.

Left-click, release, and move the cursor to **Sketch**, then **Three Point Arc** in the drop-down menu.

Left-click, release, then move the cursor down onto the grid. Left-click twice.

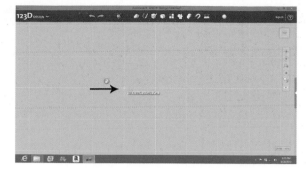

Move the cursor horizontally to the right—be careful to move straight, not up or down. A high-lighted box will appear. Type 40.

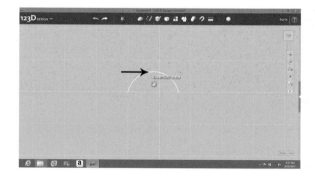

Left-click, release, and move the cursor up about 3 grid marks (15 mm).

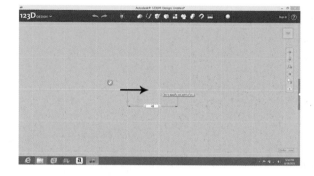

Left-click and release. Move the cursor to **Sketch** and then drop down to **Polyline**.

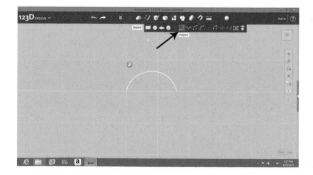

Left-click, release, and move the cursor to the left end of the arc. Left-click twice.

Move the cursor to the right until it touches the right end of the arc. The number in the highlighted box should read "40.000 mm."

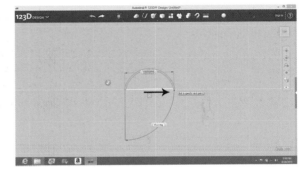

Left-click, release, and move the cursor to **Construct**. Drop down to **Extrude**.

Left-click, release, and move the cursor inside the half circle.

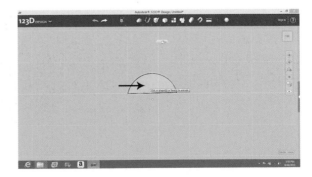

Left-click and release. Type 60 and it will appear in the box.

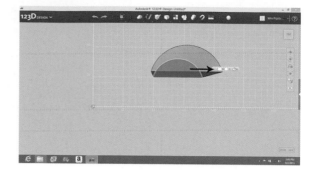

Press **Enter** and the box will go away.

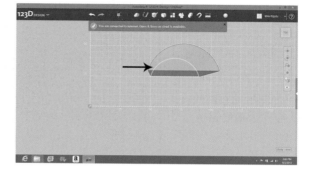

Move the cursor toward the cube in the upper right corner. A little house will appear on the upper left part of the cube when the cursor approaches.

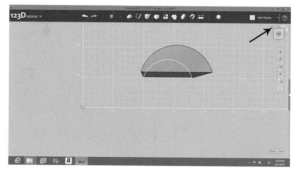

Left-click on the little house and release. Move the cursor to the part of the cube that says **Front**.

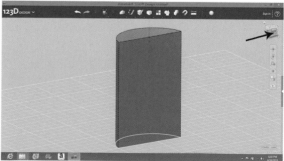

Left-click, release, and move the cursor to **Sketch**, then drop down to **Rectangle**.

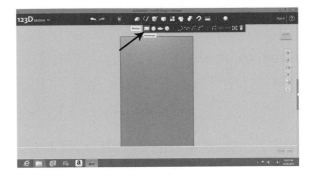

Left-click, release, and move the cursor inside the rectangle, to a spot about 6 mm right and 6 mm up from the bottom left corner. Left-click twice.

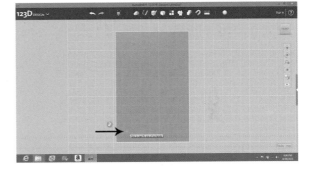

Move the cursor toward the upper right corner. The upper box should be highlighted. Type 48. Press the **Tab** key. Type 28 and it should appear in the lower box. The hole you made in the box was 50 mm by 30 mm; this insert is 48 mm by 28 mm so it will easily fit inside the hole.

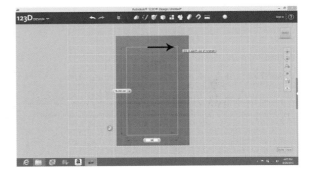

Left-click and release. Move the cursor to **Construct**, then drop down to **Extrude**.

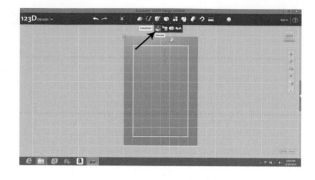

Left-click, release, and move the cursor inside the rectangle.

Left-click and release.

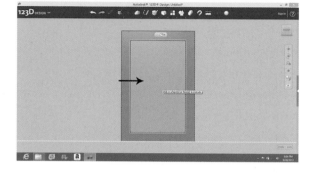

Type 5 and it should appear in the box. If you move the cursor to the right of the box, you will see the word "Merge," meaning you are adding this rectangle to your lid.

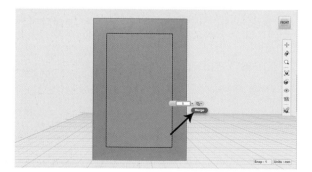

Press **Enter** and move the cursor to **123D** in the upper left corner. On the drop-down menu, select **Export STL** to save your file for printing. To save the design file, move the cursor to **123D** in the upper left corner and drop down to **Save**.

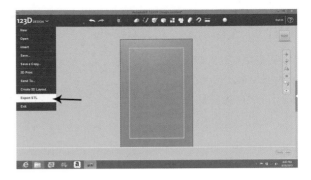

## The Oval Box

To make an oval box, you will create an oval, make it thick, and cut out an oval hole. The lid will be the same shape as the box, with an extension that is smaller than the oval box's hole.

Start a new project, go to **Primitives** and drop down to **Ellipse**.

Left-click, release, and bring the oval down onto the grid.

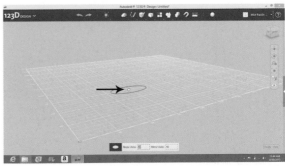

Type 30 and it will show up in the "Major Axis" box. Press the **Tab** key and type 20 for the "Minor Axis." This will make the box 60 mm long and 40 mm wide.

Left-click, release, and move the cursor to **Construct**, then drop down to **Extrude**.

Left-click, release, and move the cursor inside the oval. Left-click and release.

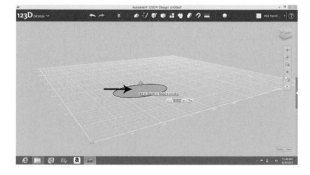

Type 35 and the oval should get thick.

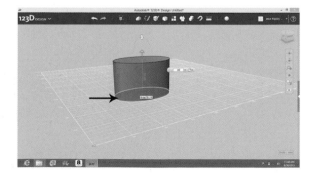

Press the **Enter** key.

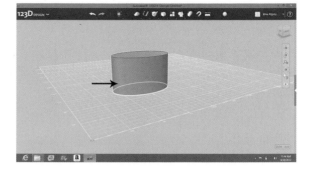

Move the cursor to the top of the little cube in the upper right corner.

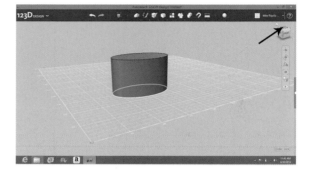

Left-click for a top view, release, and move the cursor to **Primi-tives**, then drop down to **Ellipse**.

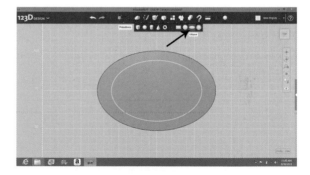

Left-click, release, and move the center of the new oval to the center of the existing oval.

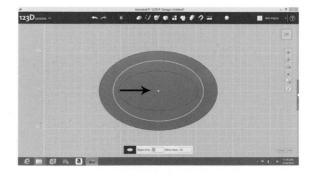

Type 25, then press the **Tab** key. Type 15. This sets the size of the hole you are going to make in this box.

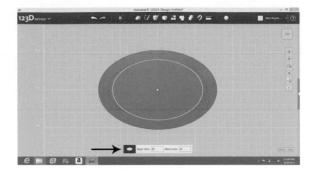

Press **Enter**.

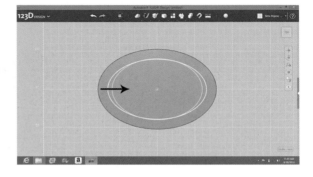

Move the cursor to **Construct** and drop down to **Extrude**.

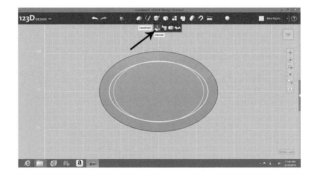

Left-click, release, then move the cursor inside the oval. Left-click and release.

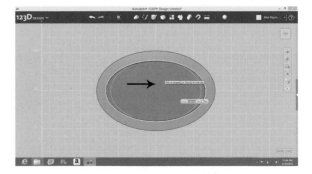

Type 30. The little box to the right of the number will say "Subtract" if you move the cursor over it.

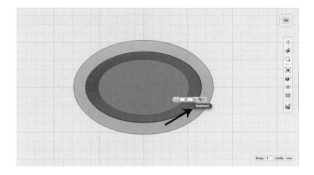

Press **Enter** and move the cursor to **123D** in the upper left corner. On the drop-down menu, move to **Export STL** to save your file for printing.

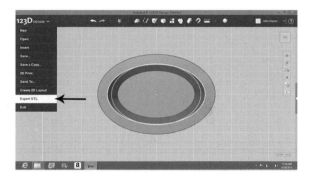

To save the design file, move the cursor to **123D** in the upper left corner and drop down to **Save**.

## Oval Box Lid

Now you will take an oval (same shape as your box) and thicken it to make a lid. You will raise an area slightly smaller than the hole in the oval box so the lid will fit nicely.

Start a new project and move the cursor to **Primitives** and drop down to **Ellipse**.

Left-click, release, then move the oval down onto the grid.

Type 30, press the **Tab** key, then type 20.

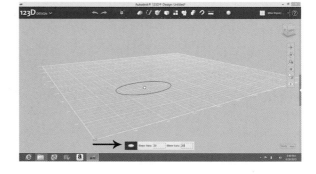

Press **Enter**, then move the cursor to **Construct**, and drop down to **Extrude**.

Left-click, release, and move the cursor inside the oval. Left-click and release.

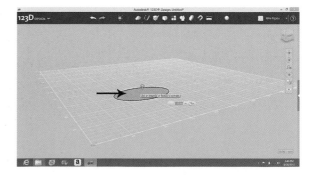

Type 5.

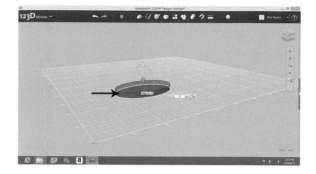

Press **Enter**, then move the cursor to the top of the little cube in the upper right corner.

Left-click, release, then move the cursor to **Primitives** and drop down to **Ellipse**.

Left-click, release, and move the new oval to the center of the existing oval.

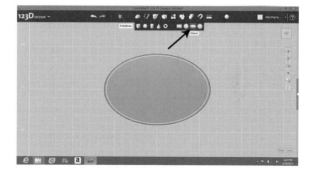

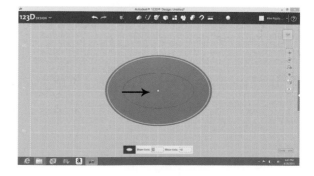

Type 24, then press the Tab key. Type 14. This makes the "lip" of your lid (24 mm by 14 mm) slightly smaller than your hole (25 mm by 15 mm when you made the oval box).

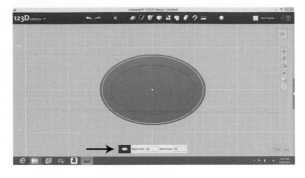

Press the **Enter** key, then move the cursor to **Construct** and drop down to **Extrude**.

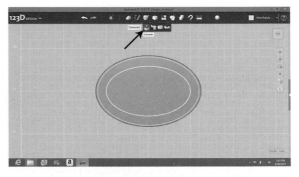

Left-click, release, and move the cursor inside the oval. Left-click and release.

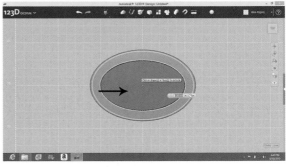

Type −5 to move the ellipse upward. If you move the cursor to the right of the box, you will see the word "Merge," meaning that you are adding to the lid. Before you press

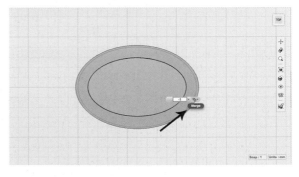

**Enter**, it's a good idea to check and see whether you are "merging" or "subtracting," since the use of positive and negative numbers in the box may not always yield the result you expect.

Press **Enter** and move the cursor to **123D** in the upper left corner. On the drop-down menu, select **Export STL** to save your file for printing.

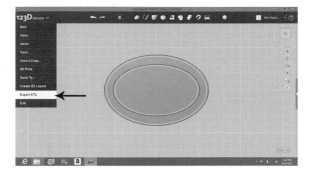

To save the design file, move the cursor to **123D** in the upper left corner and drop down to **Save**.

If you want to make nesting boxes, just use the printer's enlarge/reduce function to make different boxes. These were printed at 22, 36, 60, 100, and 165 percent.

## MAKING BOXES USING SKETCHUP

### Pirate's Box

To create the pirate's box, go to the top menu, **Camera**, and drop down to **Standard Views**. Choose **Top** from the **Standard Views**. Select the **Shape** icon (fifth from the left), then drop down to **Rectangle**. Left-click and release. Move the cursor to the center of the screen. Left-click, hold, and move the icon. In the lower right corner is a box labeled "Dimensions." Type 60,40 (you don't need to move the cursor) and press the **Return** key. Release the left-click.

Go to the **Zoom** icon (sixth from right) and left-click. Move it into the work area, and scroll the mouse until the rectangle fills about one-third of the screen. Go to the **Push/Pull** icon (sixth from left), left-click, and release. Move the icon into the upper third of the rectangle. Left-click and hold while pulling the mouse downward. Type 35 and it will appear in the "Distance" box. Press the **Enter** key. Release the left-click on the mouse.

Go to the **Tape Measure** icon (11th from the right), left-click, and release. Move the pointer to the lower left corner, then left-click and release. Move the cursor slightly to the right. Type 5, then press the **Return** key. Move the pointer from the tape measure to the little dot you just created. Left-click and release. Start moving the tape measure straight up. Type 5 and press **Return**. This will leave a dotted line and a small dot near the lower left corner.

Move the cursor to the **Shape/Rectangle** icon, then left-click and release. Move the blue dot from the cursor onto the small dot you created in the last step. Left-click, hold, and move the cursor to the right and upward slightly. Type 50,30 and press **Enter**. Release the left-click. This is the area of the "hole" that will be carved out of the block.

Move the cursor to the **Push/Pull** icon (sixth from left), left-click, and release. Pull this cursor down to the upper third of the smaller rectangle. The rectangle will fill with small black dots. Left-click and hold while pulling the cursor slightly downward. Type 30 and press the **Return** key. Release the left-click. Move the cursor to the **Select** icon (first on left), left-click, and release.

Go to the top menu, **Edit**, and drop down to **Select All**. Left-click and release.

Go to the menu at the top and choose **File**, then drop down to **Export STL**. Left-click. This file can be used to print the pirate's box.

To save the design file, choose **File**, then **Save**.

## Pirate's Box Lid

Open a new file. Go to **Camera** and drop down to **Standard Views**. Choose **Right**, left-click, and release. Move the cursor to **Arcs** (fourth from left). Select **2 Point Arc** from the drop-down menu. Left-click and release. Move the cursor to the intersection of the lines. Left-click, hold, and move the cursor to the right. Type 40 (it will show up in the box at the lower right) and press the **Return** key. Release the left-click. Move the cursor upward. Type 15 and press the **Return** key.

Use the **Zoom** tool (sixth icon from right) and scroll to make the arc cover about one-third of the screen. Move the cursor to the **Lines** icon (third from left), then select **Line** from the drop-down menu. Left-click and release. Move the cursor to the left end of the arc. The dot will turn green when it is at the endpoint of the arc—left-click and release when the dot is green. Move the cursor to the right until the green dot appears at the right end of the arc. Left-click and release.

Move the cursor to the **Push/Pull** icon (sixth from left), left-click, and release. Bring the cursor into the arc. The arc should fill with small black dots. Left-click, hold, and pull the cursor down. Type 60 and press **Enter**. Release the left-click.

Go to the top menu, **Camera**, and drop down to **Standard Views**. Choose **Bottom**, then left-click and release.

Select the **Tape Measure** icon (11th from right), left-click, and release. Take the cursor toward the lower left corner. When the little dot turns green, left-click and release. Move the cursor toward the right. Type 6 and press **Enter**. Move the dot on the tape measure over the dot you just created. Left-click and release. Pull the tape measure upward. Type 6 and press **Enter**.

Select the **Shape/Rectangle** icon (fifth from the left) by left-clicking once and releasing. Move the blue guide point until it latches onto the measured point you just created. (The blue point will turn black.) Left-click, hold, and pull the cursor right and upward. Type 48,28 and press **Enter**. Release the left-click.

Go to the **Push/Pull** icon (sixth from left), left-click, and release. Bring the cursor inside the inner rectangle (toward the top). Left-click, hold, and pull upward. Type 5 and press **Enter**. Release the left-click.

Go to the **Select** icon (first on left), left-click, and release. Go to **Edit**, then drop down to **Select All**. Left-click and release.

Go to the menu at the top and choose **File**, then drop down to **Export STL**. Left-click. This file can be used to print the lid of the pirate's box.

To save the design file, choose **File**, then **Save**.

## Oval Box

Start with a new file. Select the **Shape/Circle** icon (fifth from left), left-click, and release. Bring the circle to the intersection of the red, blue, and green lines. The center dot should turn yellow. Left-click, hold, and pull the cursor outward. Type 20 and press **Enter**. Release the left-click.

Go to the **Zoom** icon (sixth from right), left-click, release, and move the icon to the tiny circle you just created. Scroll the mouse until the circle fills about 30 percent of the screen.

Go to the **Push/Pull** icon (sixth from left), left-click, release, and move the cursor into the circle. Left-click, hold, and pull the cursor down slightly. Type 35, then press **Enter**. Release the left-click.

Select the **Shape/Circle** icon (fifth from left), left-click, release, and bring the dot to the origin (intersection of red, blue, and green lines). The dot will be yellow. Left-click, hold, and move the cursor slightly. Type 15 and press **Enter**. Release the left-click.

Go to the **Push/Pull** icon (sixth from left), left-click, release, and move the cursor into the circle. Left-click, hold, and pull the cursor down slightly. Type 30, then press **Enter**. Release the left-click.

Go to the **Select** icon (first on the left), left-click, and release. Go to the **Edit** menu, drop down to **Select All**, then left-click and release.

Go to the **Camera** menu and drop down to **Standard Views**. Move over to **Top**, then left-click and release.

Go to the **Scale** icon (10th from the left), left-click, and release. Move the pointer to the yellow line on the right. Slowly come down with the cursor until the small green square turns red. Left-click, hold, and move the cursor slightly to the right. Type 1.5,1 and press **Enter**. Release the left-click.

Go to the menu at the top and choose **File**, then drop down to **Export STL**. Left-click. This file can be used to print the oval box.

To save the design file, choose **File**, then **Save**.

## Oval Box Lid

Start with a new file. Select the **Shape/Circle** icon (fifth from left), left-click, and release. Bring the circle to the intersection of the red, blue, and green lines. The center dot should turn yellow. Left-click, hold, and pull the cursor outward. Type 20 and press **Enter**. Release the left-click.

Go to the **Zoom** icon (sixth from right), left-click, release, and move the icon to the tiny circle you just created. Scroll the mouse until the circle fills about 30 percent of the screen.

Go to the **Push/Pull** icon (sixth from left), left-click, release, and move the cursor into the circle. Left-click, hold, and pull the cursor down slightly. Type 5, then press **Enter**. Release the left-click.

Select the **Shape/Circle** icon (fifth from left), left-click, release, and bring the dot to the origin (intersection of red, blue, and green lines). The dot will be yellow. Left-click, hold, and move the cursor slightly. Type 14 and press **Enter**. Release the left-click.

Go to the **Push/Pull** icon (sixth from left), left-click, release, and move the cursor into the circle. Left-click, hold, and pull the cursor up slightly. Type 5, then press **Enter**. Release the left-click.

Go to the **Select** icon (first on the left), left-click, and release. Go to the **Edit** menu, drop down to **Select All**, then left-click and release.

Go to the **Camera** menu and drop down to **Standard Views**. Move over to **Top**, then left-click and release.

Go to the **Scale** icon (10th from the left), left-click, and release. Move the pointer to the yellow line on the right. Slowly come down with the cursor until the small green square turns red. Left-click, hold, and move the cursor slightly to the right. Type 1.5,1 and press **Enter**. Release the left-click.

Go to the menu at the top and choose **File**, then drop down to **Export STL**. Left-click. This file can be used to print the oval box lid.

To save the design file, choose **File**, then **Save**.

# 5

# Baking Powder Submarine

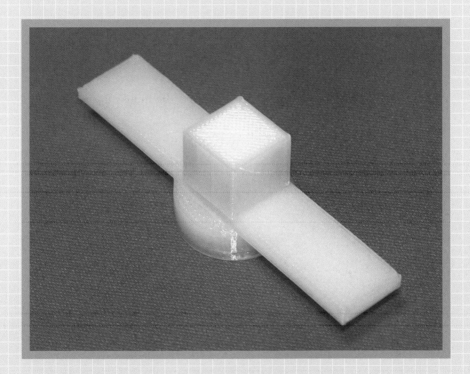

In this chapter, you will design a submarine that rises and sinks due to the reaction of baking powder (*not* baking soda) and water.

The submarine is made in two pieces. The lower part of the submarine is heavier than water. The upper tower is lighter than water, so the submarine tends to stay upright underwater. Water reacts with the baking powder and creates gas bubbles that are trapped under the submarine, causing it to rise. At the surface, the submarine becomes unstable and turns over, releasing the gas bubbles and starting the process again.

## Submarine Base

Open a new project
and move the cursor
to **Top** on the view
cube in the upper right
corner.

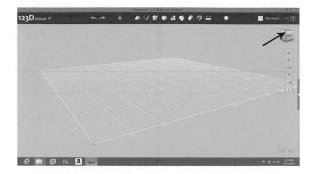

Left-click and release.
Move the cursor to
**Sketch** and drop down
to **Rectangle**.

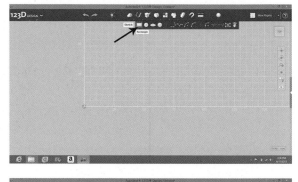

Left-click, release,
and move the cursor
toward the lower left
part of the grid.

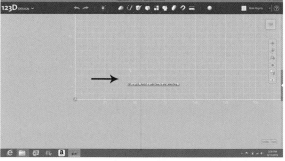

Left-click, release, and
move the cursor to the
intersection of grid
lines (indicated by a
small box).

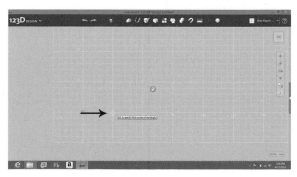

Left-click and release. Move the cursor slightly right and upward.

Type 80 and it will appear in the lower box. Press the **Tab** key and type 15.

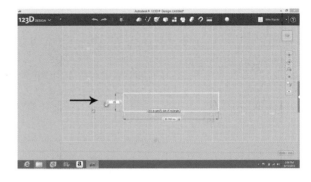

Left-click and release. Move the cursor to **Construct** and drop down to **Extrude**.

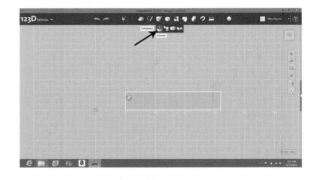

Left-click, release, and move the cursor inside the rectangle.

Left-click and release.

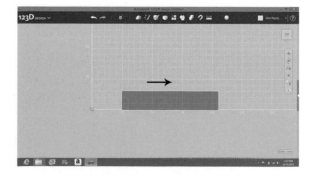

Type −3. This causes the object to be built away from you— downward.

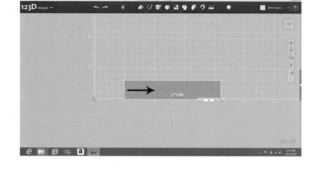

Press the **Enter** key, then move the cursor to the **Zoom** icon on the right.

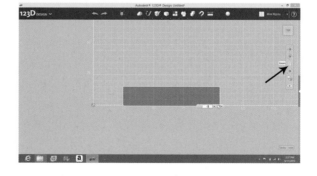

Left-click, release, then move the cursor just above the rectangle.

Scroll until the rectangle becomes six cubes high.

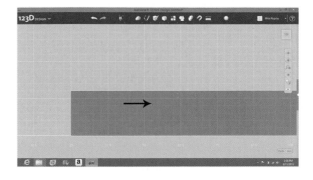

Move the cursor to **Sketch** and drop down to **Circle**.

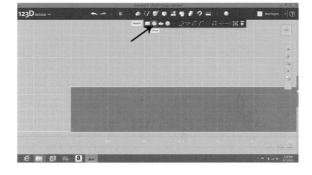

Left-click, release, then move the cursor inside the rectangle.

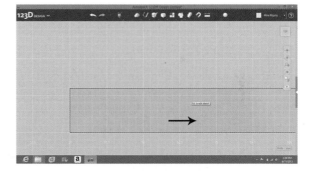

Left-click and release. Move the cursor to the center of the rectangle (16 squares from the left, 3 squares up).

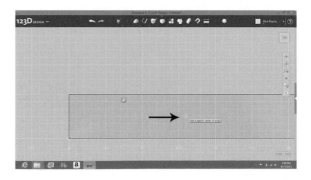

Left-click, release, and move the cursor slightly.

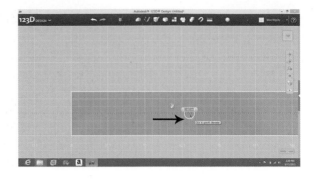

Type 2 4 and it will appear in the box. The circle will also become larger.

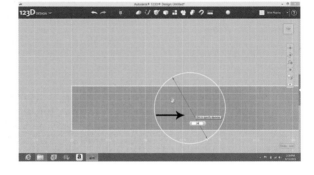

Left-click and release. Move the cursor to **Construct** and drop down to **Extrude**.

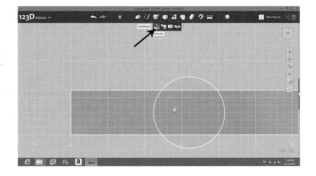

Left-click and release. Move the cursor to the lower part of the circle (below the rectangle).

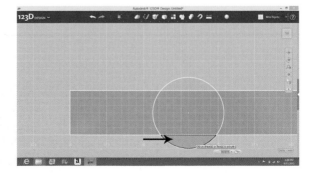

Left-click and release. Move the cursor to the center part of the circle. Left-click and release.

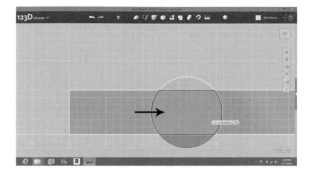

Move the cursor to the upper part of the circle. Left-click and release.

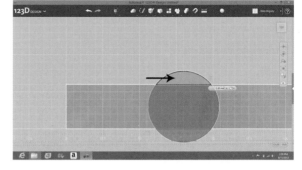

Move the cursor inside the small box (at the right of the circle). Left-click and release. Hit **Backspace** until the box is empty.

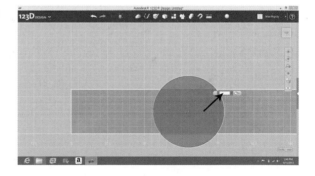

Type 8.

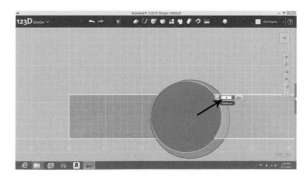

Press the **Enter** key.

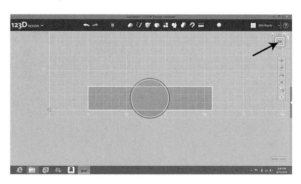

Move the cursor to the cube in the upper right corner, on the word "Top."

Left-click and release. You are doing this so the view is dead on from the top; other-wise, you would have trouble finding the center of the circle.

Move the cursor to **Sketch** and drop down to **Circle**.

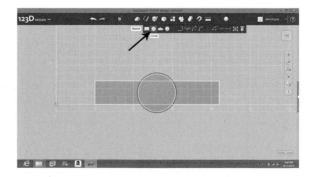

Left-click, release, and move the cursor within the circle.

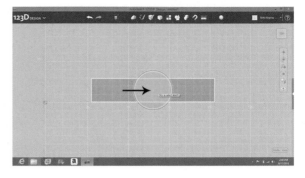

Left-click and release. Move the cursor to the center of the circle. (A small white dot is there.) Left-click, release, and move the cursor slightly.

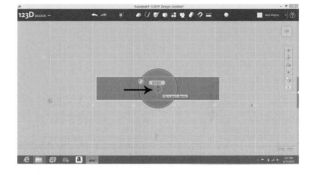

Type 7.

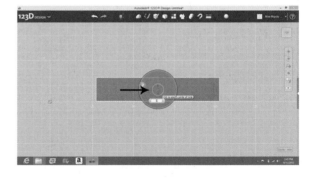

Left-click and release. Move the cursor to **Construct** and drop down to **Extrude**.

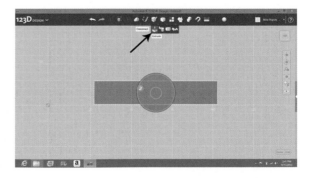

Left-click, release, then move the cursor down to the center circle. Left-click and release.

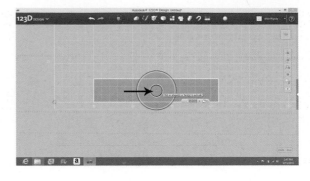

Type −11 to create a hole.

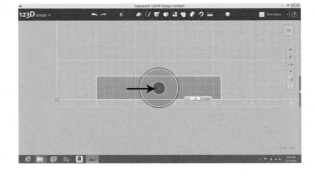

Press **Enter**. Move the cursor to **Sketch** and drop down to **Circle**.

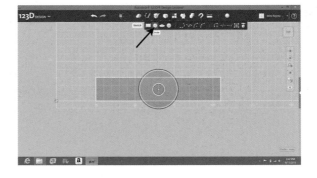

Left-click and release. Move the cursor inside the center circle.

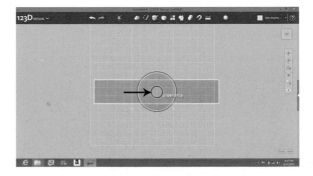

Left-click and release. Move the cursor until it "latches" onto the center. (A small circle will appear around the white center dot.)

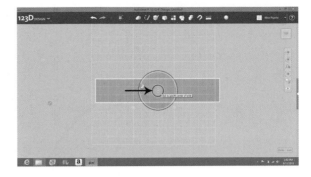

Left-click and release. Pull the cursor slightly away from the center.

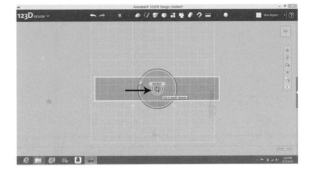

Type 10.

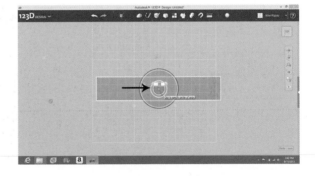

Left-click and release. Move the cursor to **Construct** and drop down to **Extrude**. Left-click and release.

Move the cursor to the ring you just created. Left-click and release.

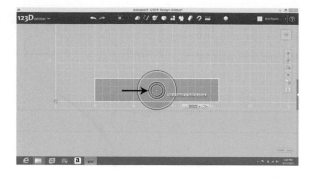

Type −8. This will make a hole 10 mm in diameter and 8 mm deep (so it doesn't go all the way through the material).

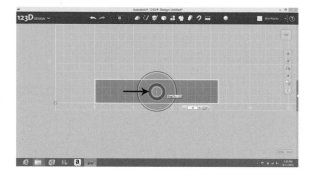

Press **Enter**, then move the cursor to **Sketch** and drop down to **Circle**. You are making concentric holes, each one wider but less deep than the one before.

Left-click and release. Move the cursor to the center circle. Left-click and release. Move the cursor until it "latches" onto the white dot in the center.

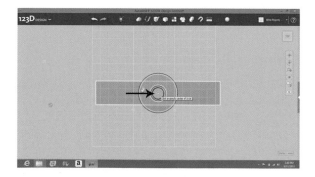

Left-click and release.
Pull the cursor slightly
away from the center.
Type 15.

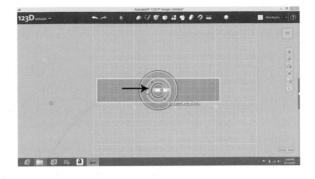

Left-click and release,
then move the cursor
to **Construct** and drop
down to **Extrude**.

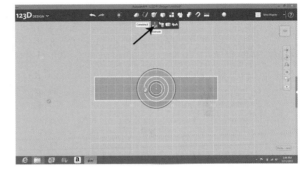

Left-click and release.
Move the cursor to the
circle you just created.
Left-click and release.

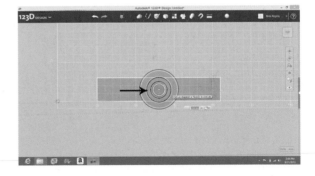

Type -5.

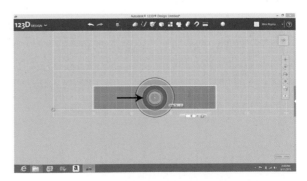

Press **Enter**, then move
the cursor to **Sketch**
and drop down to
**Circle**.

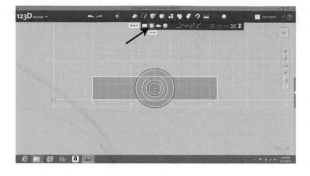

Left-click and release.
Move the cursor to the
inner circle. Left-click
and release.

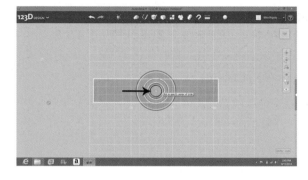

Move the cursor until
it "latches" onto the
center.

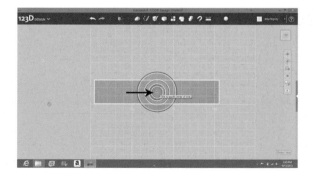

Left-click and release.
Pull the cursor slightly
away from the center.

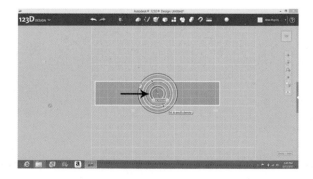

Type 22.

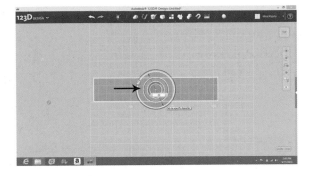

Left-click and release. Move the cursor to **Construct** and drop down to **Extrude**.

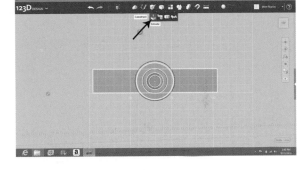

Left-click and release. Move the cursor inside the last circle you created. Left-click and release.

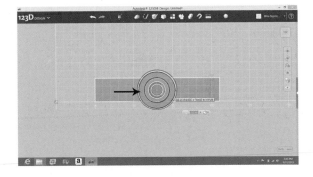

Type -2.

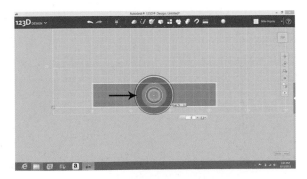

Press **Enter** and move the cursor to **123D** in the upper left corner. On the drop-down menu, find **Export STL** and save your file for printing.

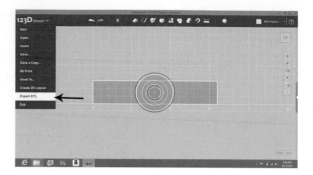

To save the design file, move the cursor to **123D** in the upper left corner and drop down to **Save**.

Print at 100 percent fill. (It needs to be solid plastic so it will sink.)

## Submarine Tower

Start a new project. Go to **Primitives** and drop down to **Box**.

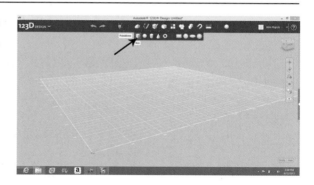

Left-click, release, and move the cursor toward the lower left part of the grid.

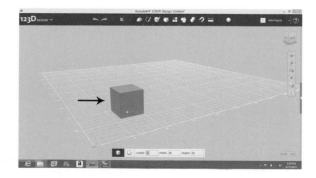

Type 15, then press the **Tab** key. Type 15, then press the **Tab** key again. Type 13.

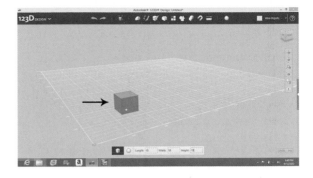

Press **Enter**. Move the cursor to the top of the cube in the upper right corner.

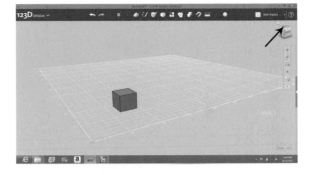

Left-click and release. Move the cursor to **Sketch** and drop down to **Rectangle**.

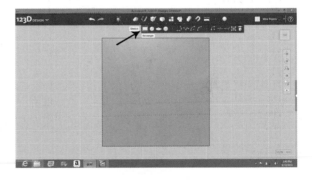

Left-click, release, and move the cursor into the square.

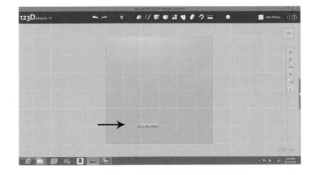

Left-click and release. Move the cursor to a position about 2 mm right and 2 mm up from the lower left corner. Left-click and release. Move the cursor slightly.

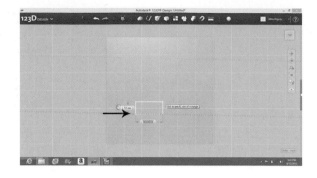

Type 7, then press the **Tab** key. Type 7.

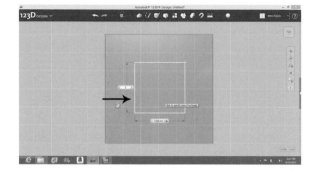

Left-click, release, and move the cursor to **Construct** and drop down to **Extrude**.

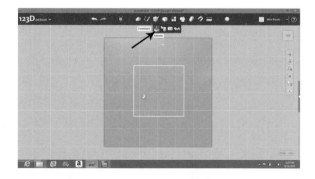

Left-click and release. Move the cursor into the smaller square, then left-click and release.

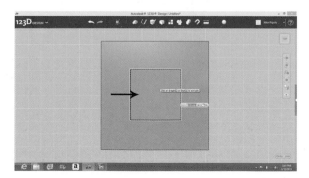

Type -11.

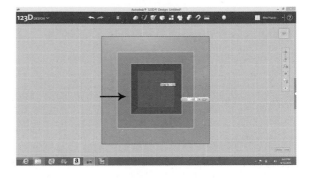

Press **Enter** and move the cursor to **123D** in the upper left corner. On the drop-down menu, select **Export STL** and save your file for printing.

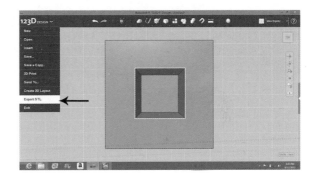

To save the design file, move the cursor to **123D** in the upper left corner and drop down to **Save**.

Print at 10 percent fill. (The walls need to be hollow on the tower—filled with air—so the submarine will stay upright when underwater.)

Set the submarine base on a flat surface, round side down. Take the tower (hole facing down) and use superglue to affix the tower to the submarine base.

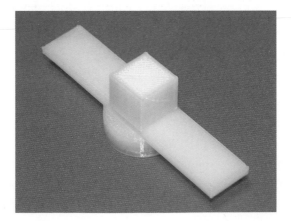

After the glue dries, turn the submarine over and insert baking powder into the tower through the hole in the bottom. Shake the powder down, and wipe away the excess with your finger.

Place underwater, hole facing down, shake, and release. After a few seconds, the submarine will surface. After a few more seconds, the submarine will turn sideways, release a gas bubble, and sink. The submarine should surface and sink 8 to 10 times before needing a new charge of baking powder. Just wash out the old powder and reload.

## MAKING A SUBMARINE USING SKETCHUP

### Submarine Base

Start with a new file. Go to the top menu **Camera** and drop down to **Standard Views**. Left-click on **Top** and release.

Go to the **Shape/Rectangle** icon. Left-click and release. Pull the cursor to the intersecting lines. Left-click, hold, and pull the cursor right and upward. Type 80,15. Press the **Enter** key. Release the left-click.

Go to the **Zoom** icon. Left-click and release. Bring the magnifying lens near the rectangle, then scroll until the rectangle fills most of the work area.

Go to the **Push/Pull** icon. Left-click and release. Bring the cursor into the rectangle. The rectangle should fill with small black dots. Left-click and hold. Move the cursor upward, and the rectangle should turn white. (The dots go away.) If the dots remain, move the cursor downward until the rectangle turns white. Type 3 and press the **Enter** key. Release the left-click.

Go to the **Tape Measure** icon, left-click, and release.

Move the blue part of the icon to the lower left corner of the rectangle. It will turn into a green circle. Left-click, hold, and move the cursor toward the right. Type 40 and press the **Enter** key. Release the left-click.

Move the cursor until it latches onto the spot you just made. The words "Guide Point" will appear briefly. Left-click and hold. Move the cursor upward. Type 7.5 and press the **Enter** key. Release the left-click. You now have a guide mark in the center of the rectangle (40 mm from the left, 7.5 mm up).

You are now going to make a series of circles, each of which will be a different width and thickness. Move the cursor to the **Shape/Circle** icon. Left-click and release.

Move the cursor to the center guide point you just created. Left-click and hold. Pull the cursor outward. Type 3.5 and press the **Enter** key. Release the left-click.

Move the cursor to the center guide point. Left-click and hold. Pull the cursor outward. Type 5 and press **Enter**. Release the left-click.

Move the cursor to the center guide point. Left-click and hold. Pull the cursor outward. Type 7.5 and press **Enter**. Release the left-click.

Move the cursor to the center guide point. Left-click and hold. Pull the cursor outward. Type 11 and press **Enter**. Release the left-click.

Move the cursor to the center guide point. Left-click and hold. Pull the cursor outward. Type 12 and press **Enter**. Release the left-click.

Move the cursor up to the **Push/Pull** icon near the top of the page. Left-click and release.

Move the cursor to the left part of the outer ring (the rectangle will divide the ring into four separate sections). Left-click and hold. Move the cursor until it pulls the left ring segment upward. Type 8 and press **Enter**. Release the left-click. Repeat with the other three sections of the outer ring.

Move the cursor to the left part of the next ring inward (also divided into four sections). Left-click and hold. Move the cursor until it pulls the ring segment upward. Type 6 and press **Enter**. Release the left-click. Repeat with the other three segments.

Move the cursor to the next ring inward, which is all in one piece. Left-click and hold. Move the cursor until it pulls the ring segment upward. Type 3 and press **Enter**. Release the left-click.

Move the cursor to the center circle (two steps inward). Left-click and hold. Move the cursor upward until the tiny black dots disappear. Type 1 and press **Enter**. Release the left-click.

Move the cursor to the **Select** icon (first on the left). Left-click and release. Go to the **Edit** menu and drop down to **Select All**. Left-click and release.

Go to the menu at the top and choose **File**, then drop down to **Export STL**. Left-click. This file can be used to print the submarine base.

To save the design file, choose **File**, then **Save**.

## Submarine Tower

Start with a new file. Go to the top menu **Camera** and drop down to **Standard Views**. Left-click on **Top** and release.

Go to the **Shape/Rectangle** icon. Left-click and release. Pull the cursor to the intersecting lines. Left-click, hold, and pull the cursor right and upward. Type 15,15. Press the **Enter** key. Release the left-click.

Go to the **Zoom** icon. Left-click and release. Bring the magnifying lens near the rectangle, then scroll until the rectangle fills most of the work area.

Go to the **Push/Pull** icon. Left-click and release. Bring the cursor into the rectangle. The rectangle should fill with small black dots. Left-click and hold. Move the cursor upward, and the rectangle should turn white. (The dots go away.) If the dots remain, move the cursor downward until the rectangle turns white. Type 13 and press the **Enter** key. Release the left-click.

Go to the **Tape Measure** icon, left-click, and release.

Move the blue part of the icon to the lower left corner of the rectangle. It will turn into a green circle. Left-click, hold, and move the cursor toward the right. Type 4 and press the **Enter** key. Release the left-click.

Move the cursor until it latches onto the spot you just made. The words "Guide Point" will appear briefly. Left-click and hold. Move the cursor upward. Type 4 and press the **Enter** key. Release the left-click.

Move the cursor to the **Rectangle** icon. Left-click and release.

Pull the cursor down until the point latches onto the guide point you created in the last step.

Left-click and hold. Pull the cursor right and upward. Type 7 , 7 and press **Enter**. Release the left-click.

Move the cursor to the **Push/Pull** icon. Left-click and release.

Bring the cursor into the lower part of the center square. Left-click and hold. Move the cursor upward. Type 11 and press **Enter**. (The center square should get smaller and be surrounded on three sides by a dark gray color.)

Move the cursor to the **Select** icon (first on the left). Left-click and release. Go to the **Edit** menu and drop down to **Select All**. Left-click and release.

Go to the menu at the top and choose **File**, then drop down to **Export STL**. Left-click. This file can be used to print the submarine tower.

To save the design file, choose **File**, then **Save**.

# 6

# Bicycle Spinner

In this chapter you will make a snap-on windmill for bicycle handlebars. The faster you go, the more it will turn and the more noise it will make.

You'll start by taking a rectangular piece of plastic and cutting out a hole the size of a bicycle's handlebar. Next, you'll create a propeller and a washer to hold the propeller in place.

## The Spinner Bracket

Start a new project. Go to **Primitives** and drop down to **Box**.

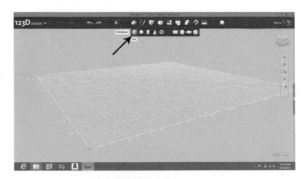

Left-click, release, and drop down to the lower left part of the work area. Type 40, then press the **Tab** key. Type 100, then press the **Tab** key, followed by 15.

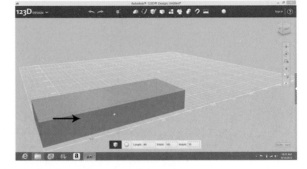

Press the **Enter** key, then move the cursor to the top of the cube in the upper right corner.

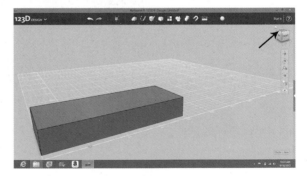

Left-click, release, and move the cursor to **Sketch** and drop down to **Circle**.

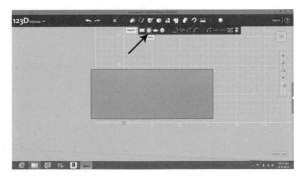

Left-click, release, and move the cursor inside the rectangle.

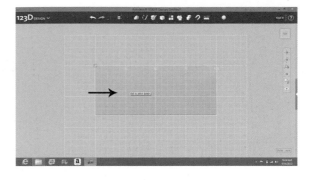

Left-click and release. Move the cursor to a point four squares right and four squares up from the lower left corner. Left-click, release, and move the cursor slightly.

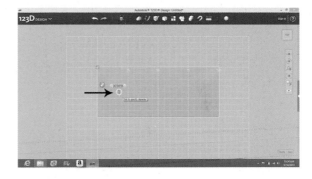

Type 22. This is the diameter (in millimeters) of a typical handlebar. If you know yours is larger or smaller, type the appropriate number here instead of 22.

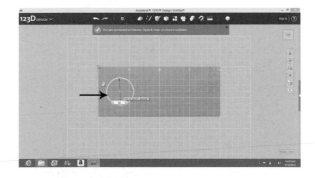

Left-click, release, move the cursor to **Construct**, and drop down to **Extrude**. Left-click and release.

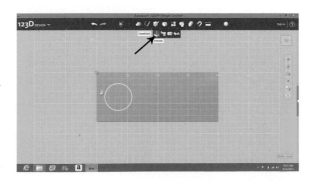

Move the cursor inside the circle.

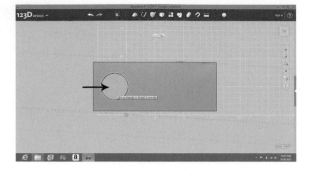

Left-click and release. Type −15 to create a hole.

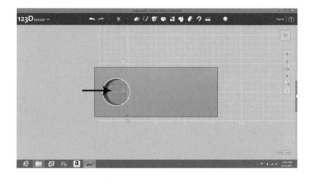

Press the **Enter** key.

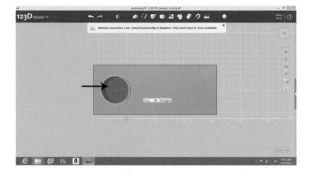

Move the cursor to **Sketch** and drop down to **Polyline**.

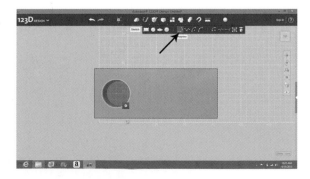

Left-click, release, and move the cursor inside the circle.

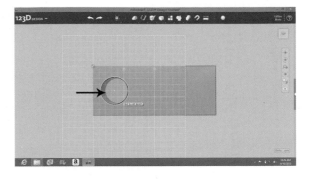

Left-click and release. Move the cursor to a point just inside the circle as shown. Left-click and release.

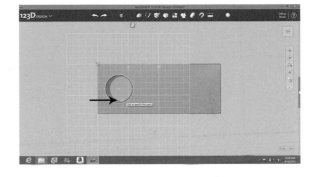

Pull the cursor horizontally to the right, until it is outside the rectangle. Left-click and release.

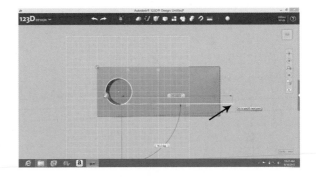

Move the cursor straight up until it is above the top of the rectangle. Left-click and release.

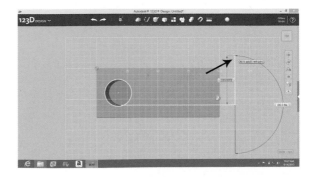

Move the cursor horizontally left until you reach the midpoint of the rectangle (50 mm). Left-click and release.

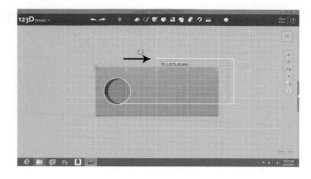

Move the cursor down until you reach the second grid mark inside the rectangle. Left-click and release.

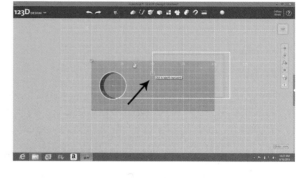

Move the cursor left horizontally until it is inside the circle. Left-click and release.

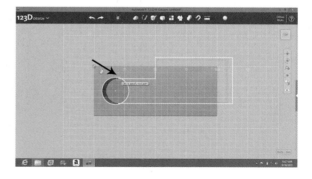

Move the cursor down until it connects with the first point you created. Left-click and release.

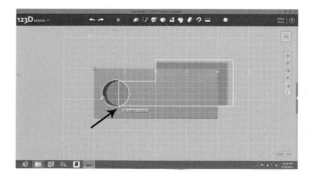

Move the cursor to **Construct** and drop down to **Extrude**.

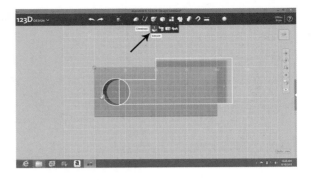

Left-click, release, and move the cursor inside the area you just created. Left-click and release.

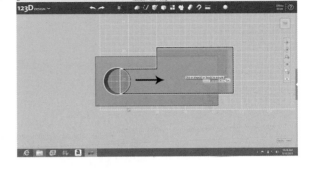

Type −15 to remove the unwanted material.

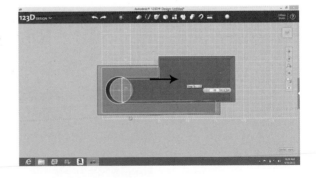

Press **Enter** and move the cursor to the cube in the upper right corner (the lower part of the cube).

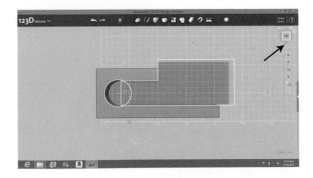

Left-click, release, move the cursor to **Sketch**, and drop down to **Circle**.

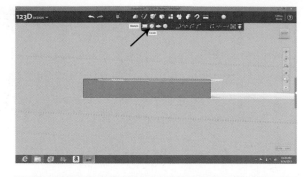

Left-click and release. Move the cursor inside the rectangle.

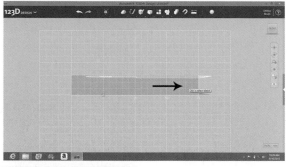

Left-click and release.

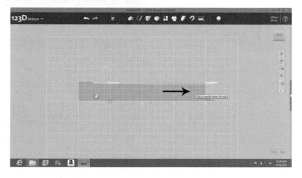

Move the cursor to a point about 1½ grid marks up, 2 grid marks left of the right end of the rectangle. (Approximate measures will work fine in this situation.) Left-click, release, and move the cursor slightly. Type 4.

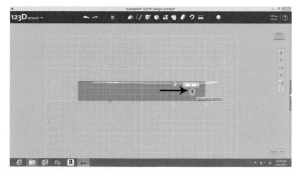

Left-click, release, and move the cursor to **Construct**, then drop down to **Extrude**.

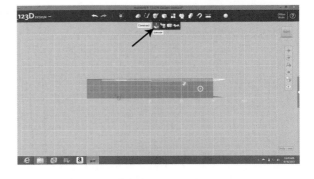

Left-click, release, and move the cursor inside the small circle.

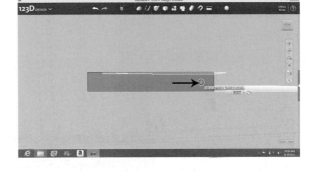

Left-click, then release. Type 30. (This specifies how far out the propeller shaft will extend.)

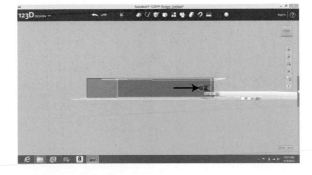

Press **Enter** and move the cursor to **123D** in the upper left corner. On the drop-down menu, **Export STL** and save your file for printing.

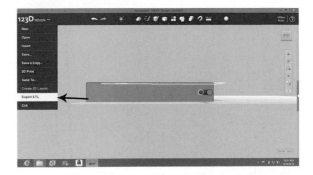

To save the design file, move the cursor to **123D** in the upper left corner and drop down to **Save**.

## Propeller

To make a propeller, you will take a box and make a hole in the middle so it will fit on a shaft. You will then go to one end and carve away the material that "isn't a propeller." You will do the same at the other end.

Start a new project, then go to **Primitives**, and drop down to **Box**.

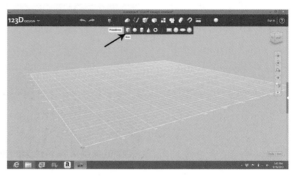

Left-click, release, and move the cursor down to the lower left area of the screen. Press the **Tab** key. Type 100 for the width. (Keep the default value, 20, for the other two dimensions.)

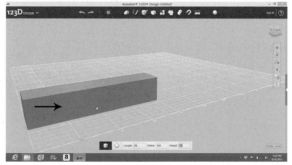

Press the Enter key. This creates a box that is 20 mm high, 100 mm wide, and 20 mm deep. Move the cursor to the small cube in the upper right of the screen. Place the cursor on the word "Front."

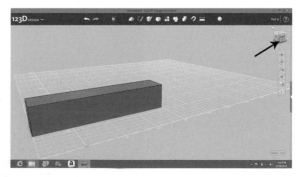

Left-click and release. Move the cursor to **Sketch**, then drop down to **Circle**.

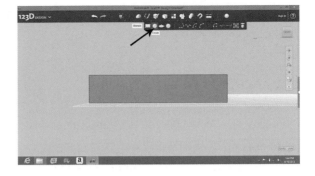

Left-click and release. Move the cursor to the center of the rectangle. Left-click two times, then move the cursor slightly.

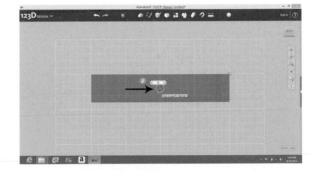

Type 6.

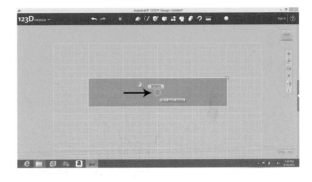

Left-click, release, then move to **Construct** and drop down to **Extrude**.

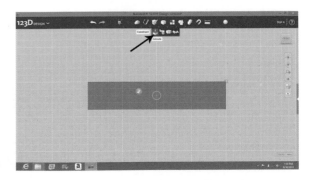

Left-click and release.
Move the cursor to the
center of the circle.
Left-click and release.

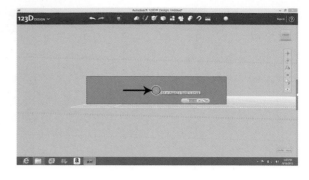

Type -20 to create
a hole.

Press the **Enter** key.
Move the cursor to
the cube in the upper
right corner. Place the
cursor on the arrow to
the right of the word
"Front."

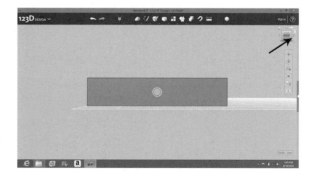

Left-click and release.
Move the cursor to
**Sketch** and drop down
to **Polyline**.

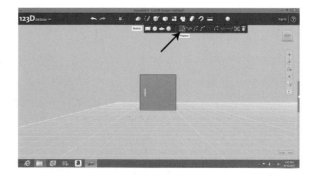

Left-click and release. Move the cursor inside the square. Left-click and release. Move the cursor to the right of the square's upper left corner (about ½ of a grid mark) right along the edge. Left-click and release.

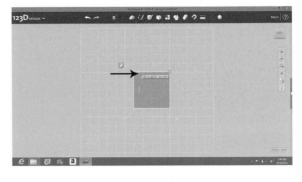

Bring the cursor down and to the right until it is outside the box (about 135 degrees—a "degree" indicator will appear). Left-click and release. Move the cursor up (0 degrees) until it is above

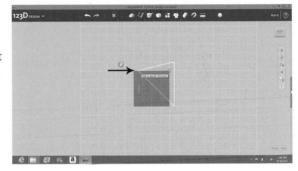

the box. Left-click and release. Move the cursor until it connects with the first point. (A small square will appear.) Left-click and release.

Move the cursor to **Construct** and drop down to **Extrude**.

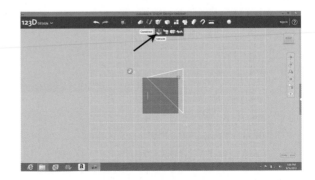

Left-click and release. Move the cursor inside the area you just created. Left-click and release. Type − 4 0. This is the depth of material you will remove.

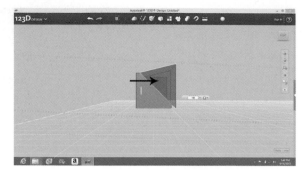

Press **Enter**, then move the cursor to **Sketch** and drop down to **Polyline**.

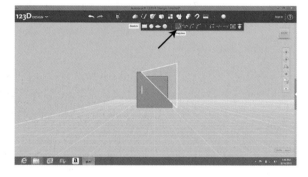

Left-click and release. Move the cursor into the square (left side). Left-click and release. Choose the first point (about ½ a grid mark below the upper left corner along the edge).

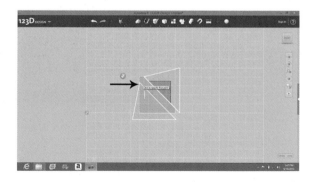

Left-click and release. Move the cursor down and right (about 135 degrees) until the cursor is below the box. Left-click and release. Move the cursor left (90 degrees) until it is left of the box. Left-click and release. Move the cursor up until it connects with the first point. Left-click and release.

Take the cursor to **Construct** and drop down to **Extrude**. Left-click and release. Move the cursor into the triangle you just created. Left-click and release. Type -40. Press **Enter**.

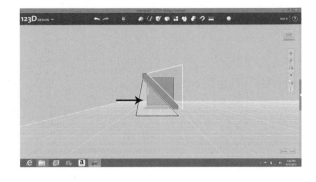

Move the cursor to the cube in the upper right, just left of the word "Right."

Left-click and release.

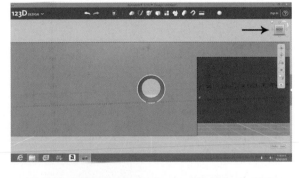

Move the cursor to **Sketch** and drop down to **Polyline**.

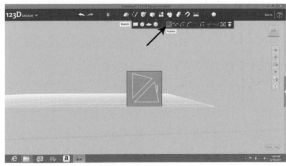

Left-click and release. Move the cursor inside the square. Left-click and release. Move the cursor ½ a grid mark to the right of the upper left corner. Left-click and release.

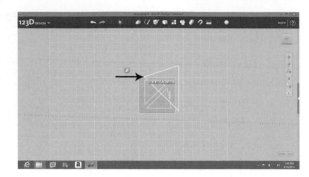

Pull the cursor down and to the right. Left-click and release. Pull the cursor straight up. Left-click and release. Pull the cursor back to the starting point. Left-click and release.

Move the cursor to **Construct** and drop down to **Extrude**. Left-click and release. Move the cursor inside the triangle you just created. Left-click and release.

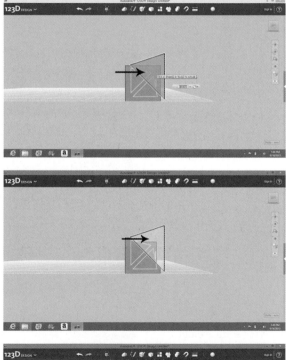

Type −40 and press **Enter**.

Move the cursor to **Sketch** and drop down to **Polyline**.

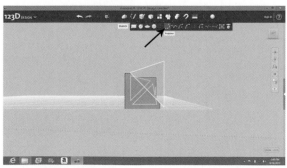

Left-click and release. Move the cursor inside the square (left side). Left-click and release. Move the cursor to the first point (½ grid mark below the upper left corner). Left-click and

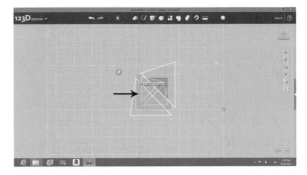

release. Move the cursor right and down (about 45 degrees) until it is below the square. Left-click and release. Move the cursor left (90 degrees) until it is left of the square. Left-click and release. Bring the cursor back to the starting point. Left-click and release.

Move the cursor to **Construct** and drop down to **Extrude**. Left-click and release. Move the cursor inside the triangle you just created. Left-click and release.

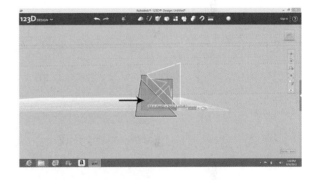

Type −40. Press **Enter** and move the cursor to **123D** in the upper left corner. On the drop-down menu, select **Export STL** and save your file for printing.

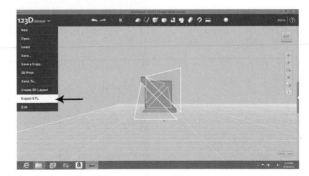

To save the design file, move the cursor to **123D** in the upper left corner and drop down to **Save**.

## The Washer

Start a new project. Go to **Primitives** and drop down to **Cylinder**.

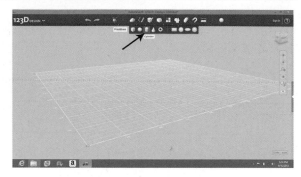

Left-click and release. Move the cursor to the lower left area on the screen. Type 5 and press the **Tab** key. Type 3.

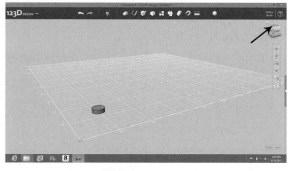

Press **Enter** and move the cursor to **Top** on the view cube.

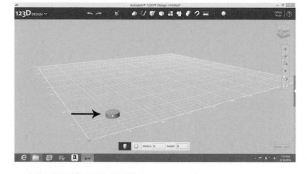

Left-click and release. Move the cursor to **Sketch** and drop down to **Circle**.

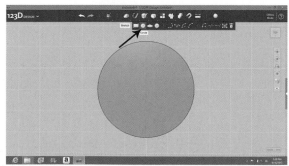

Left-click and release. Move the cursor to the center of the circle. Left-click two times, then move the cursor slightly.

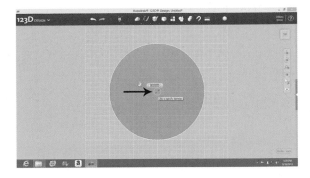

Type 5.25. Left-click and release.

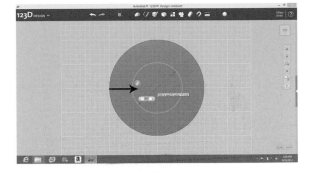

Move the cursor to **Construct** and drop down to **Extrude**.

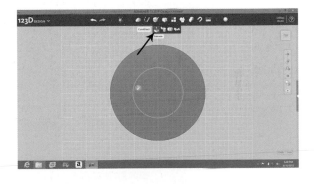

Left-click and release. Move the cursor to the center circle. Left-click and release.

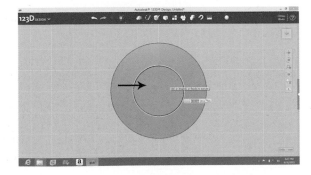

Type −3 to create a hole. Press **Enter** and move the cursor to **123D** in the upper left corner. On the drop-down menu, select **Export STL** and save your file for printing.

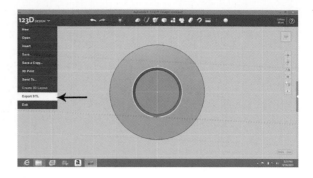

To save the design file, move the cursor to **123D** in the upper left corner and drop down to **Save**.

Print two washers, one spinner bracket, and one propeller blade.

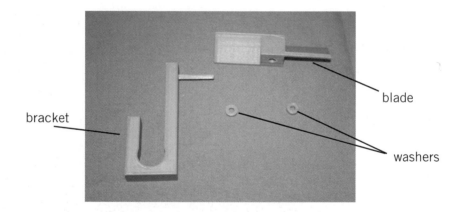

Place one washer on the bracket shaft, followed by the blade.

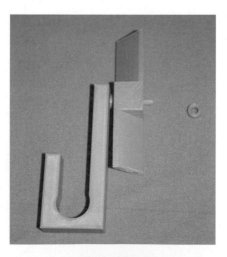

Place the second washer on the end of the bracket to hold the blade in place. This may fit by friction, but a drop of glue would help.

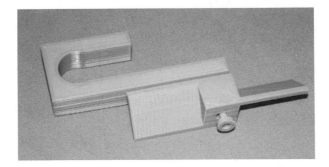

Slide the spinner on a bike's handlebar and take off.

## MAKING THE BIKE SPINNER USING SKETCHUP

## The Spinner Bracket

Go to **Camera**, then drop down to **Standard Views**. Left-click **Top** and release.

Go to the **Lines** icon (third from left), then select **Line** from the drop-down menu. Left-click and release. Bring the cursor to the intersection of the lines. Left-click, hold, and pull the cursor toward the right. Type 100 and press **Enter**. Release the left-click.

Go to the **Zoom** icon, left-click and release. Bring the magnifying lens into the work area and scroll until the black line covers about one-third of the screen.

Go back to **Lines** and drop down to **Line**. Left-click and release. The line should start where you left off (the right end of the black line). Left-click and hold. Pull the cursor upward. Type 10 and press **Enter**. Pull the cursor left. Type 75 and press **Enter**. Pull the cursor up. Type 20 followed by **Enter**. Pull the cursor right. Type 25 followed by **Enter**. Go up with the cursor. Type 10 followed by **Enter**. Pull the cursor left. Type 50 and press **Enter**. Go down with the cursor. Type 40 followed by **Enter**. An enclosed area should turn blue. Release the left-click.

Choose the **Tape Measure** icon and bring the dot to the lower left corner of the polygon just drawn. Left-click, hold, and move the cursor right. Type 20 and press **Enter**. Release the left-click.

Move the tape measure dot until it rests over the point you just made. The words "Guide Point" should appear briefly. Left-click, hold, and pull the cursor upward. Type 20 and press **Enter**. Release the left-click.

Select the **Shape/Circle** icon and bring the dot down to the guide point you just created. Left-click, hold, and pull the cursor out slightly. Type 11 and press **Enter**. Release the left-click.

Go to the **Select** icon (first on left), left-click, and release.

Bring the pointer icon down to the left part of the circle. Left-click and release. Little black dots will appear in part of the circle. Go to **Edit** and drop down to **Delete**. Left-click and release. Move the pointer icon to the right portion of the circle. Left-click and release. Go to **Edit** and drop down to **Delete**. Left-click and release.

Select the **Push/Pull** icon. Move the cursor into the blue polygon area. Left-click, hold, and pull slightly upward. Type 15 and press **Enter**. Release the left-click.

Go to **Camera** and drop down to **Standard Views**. Go to **Front**. Left-click and release.

Choose the **Tape Measure** icon and move to the lower right corner of the white rectangle. Left-click, hold, and move the cursor toward the left. Type 10 and press **Enter**. Release the left-click. Move the icon pointer until it rests over the guide point you just made. Left-click, hold, and pull the cursor upward. Type 7.5 and press **Enter**. Release the left-click.

Select the **Shape/Circle** icon and bring the pointer to the last guide point you made. Left-click, hold, and pull the cursor away from the center. Type 2 and press **Enter**. Release the left-click.

Select the **Push/Pull** tool. Bring it to the small circle. (Black dots will fill the circle.) Left-click, hold, and move the cursor. Type 30 and press **Enter**. Release the left-click.

Go to the **Select** icon. Left-click and release. Go to **Edit**, then drop down to **Select All**. Left-click and release.

Go to the menu at the top and choose **File**, then drop down to **Export STL**. Left-click. This file can be used to print the bike spinner bracket.

To save the design file, choose **File**, then **Save**.

## Propeller

Go to **Camera** and drop down to **Standard Views**. Choose **Top**, then left-click and release.

Select the **Shape/Rectangle** icon. Move the point to the intersecting lines. Left-click, hold, and move the cursor right and upward. Type 100,20 and press **Enter**. Release the left-click.

Select the **Zoom** icon. Scroll until the rectangle fills most of the screen.

Select the **Tape Measure** icon. Move the pointer to the lower left corner of the rectangle. Left-click and hold. Pull the cursor to the right. Type 50 and press **Enter**. Release the left-click.

Move the pointer to the guide point you just created. (It will briefly display the words "Guide Point.") Left-click and hold. Pull the cursor upward. Type 10 and press **Enter**. Release the left-click. You now have a guide point in the center of the rectangle.

Select the **Shape/Circle** icon. Bring it down until the center of the circle latches onto the center guide point. Left-click, hold, and pull the point away from the center. Type 3 and press **Enter**. Release the left-click.

Choose the **Select** icon. Bring it inside the circle, then left-click and release. The circle should fill with small black dots. If the circle does not fill with dots, it will be necessary to use the zoom tool and make the rectangle larger (nearly fill the screen).

Go to **Edit** and drop down to **Delete**. Left-click and release.

Select the **Push/Pull** icon. Bring it into the rectangle. Left-click and hold. Pull the cursor upward. Type 20 and press **Enter**. Release the left-click.

Go to **Camera** and drop down to **Standard Views**. Choose **Right**. Left-click and release.

Go to the **Lines** icon (third from left), then select **Line** from the drop-down menu. Left-click and release. Bring the cursor to the upper part of the square, about 1/10 of the way 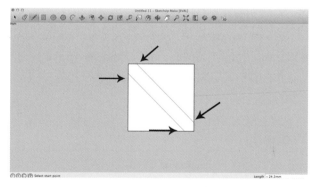 from the upper left corner to the upper right corner. Left-click, hold, and pull the line toward the lower right. Release the left-click. Go to the upper left corner and move the cursor down about 1/10 of the way to the lower left corner. Left-click, hold, and pull a line toward the lower right. Release the left-click.

Select the **Push/Pull** icon. Bring it into the rectangle, near the upper right corner. Left-click, hold, and move the cursor diagonally toward the lower left corner. (The upper right triangle should become smaller.) Type 4 0 and press **Enter**. Release the left-click.

Move the cursor toward the lower left corner (still inside the white figure). Left-click, hold, and move the cursor slightly toward the upper right. Type 4 0 and press **Enter**. Release the left-click.

Move the cursor to **Camera** and drop down to **Standard Views**. Select **Left**, then left-click and release. Repeat drawing the lines and pushing away 40 mm of material just like you did on the right end of the block.

Go to the **Select** icon. Left-click and release. Go to **Edit**, then drop down to **Select All**. Left-click and release.

Go to the menu at the top and choose **File**, then drop down to **Export STL**. Left-click. This file can be used to print the propeller.

To save the design file, choose **File**, then **Save**.

## The Washer

Start a new file. Go to **Camera** and drop down to **Standard Views**. Choose **Top**, then left-click and release.

Select the **Shape/Circle** icon and bring it to the intersection of the red and green lines. Left-click, hold, and pull the cursor away from the center. Type 5 and press **Enter**. Release the left-click.

Select the **Zoom** icon. Enlarge the circle so it fills most of the screen.

Select the **Shape/Circle** icon. Bring the pointer to the center of the circle. Left-click, hold, and move the cursor away from the center. Type 2 . 6 and press **Enter**. Release the left-click.

Move the cursor to the **Select** icon, then left-click and release. Bring the cursor inside the smaller circle. Left-click and release. (The center circle will fill with small black dots.)

Go to **Edit** and drop down to **Delete**. Left-click and release.

Select the **Push/Pull** icon. Bring it into the semi-circle. Left-click, hold, and pull the cursor down. Type 3 and press **Enter**. Release the left-click.

Go to the **Select** icon. Left-click and release. Go to **Edit**, then drop down to **Select All**. Left-click and release.

Go to the menu at the top and choose **File**, then drop down to **Export STL**. Left-click. This file can be used to print the two washers.

To save the design file, choose **File**, then **Save**.

# 7

# Dollhouse

In this chapter, you will design the pieces needed to build a modular doll house. You will start with a plain square section that can be used for a floor, wall, or ceiling. Next, you will design a square with a rectangular door opening. Then you will make a square with an oval window opening. And lastly, you will design a clip to hold all the pieces together.

## Floor, Wall, or Ceiling Pieces

Go to **Primitives** and drop down to **Box**.

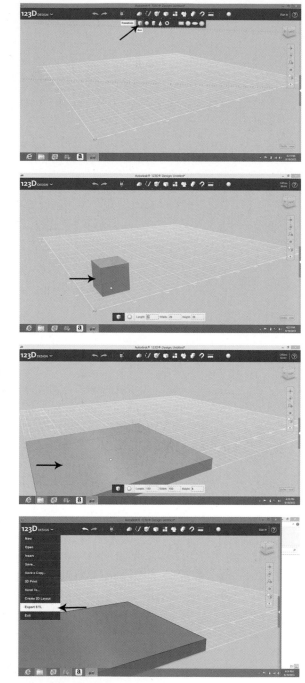

Left-click, release, and drag the box down toward the lower left part of the work area.

Type 100, then press the **Tab** key. Type 100 and press the **Tab** key. Type 8.

Press **Enter** and move the cursor to **123D** in the upper left corner. On the drop-down menu, select **Export STL** and save your file for printing.

To save the design file, move the cursor to **123D** in the upper left corner and drop down to **Save**.

# Door

Go to **Primitives** and drop down to **Box**.

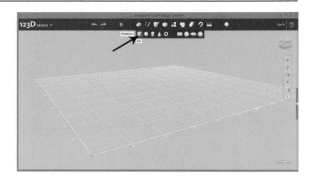

Left-click and release. Drag the box to the lower left corner.

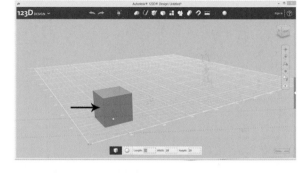

Type 100, then press the **Tab** key. Type 100 a second time, then press the **Tab** key again. Type 8.

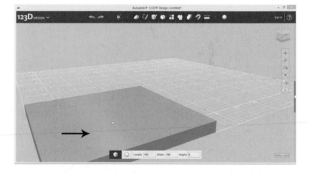

Press **Enter**, then move the cursor to **Top** on the view cube.

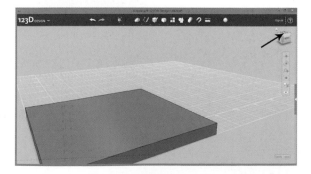

Left-click and release. Move the cursor to **Sketch** and drop down to **Rectangle**.

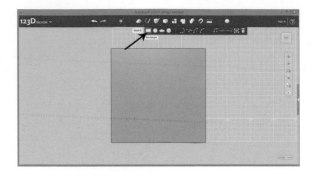

Left-click, release, and move the cursor inside the square.

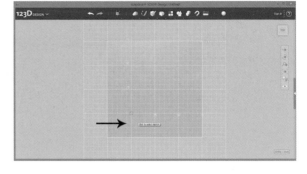

Left-click and release. Move the cursor to a point 40 mm (8 grid marks) right of the lower left corner. Left-click, release, and pull the cursor to the right and upward.

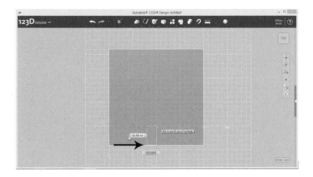

Type 35, then press the **Tab** key. Type 65.

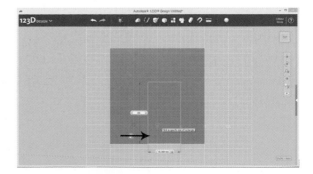

Left-click, release, and move the cursor to **Construct**, then drop down to **Extrude**.

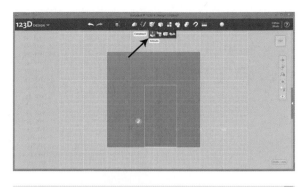

Left-click, release, and move the cursor inside the door area. Left-click and release.

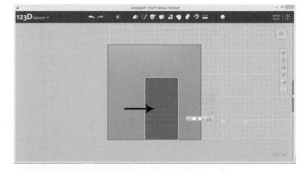

Type −8 to create the opening for the door.

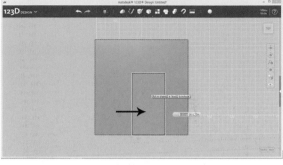

Press **Enter** and move the cursor to **123D** in the upper left corner. On the drop-down menu, find **Export STL** and save your file for printing.

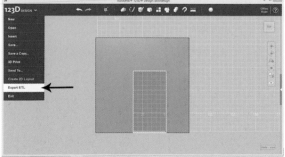

To save the design file, move the cursor to **123D** in the upper left corner and drop down to **Save**.

## Window

Go to **Primitives** and drop down to **Box**.

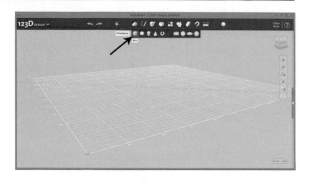

Left-click, release, and drag the box to the lower left area of the screen.

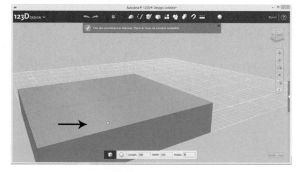

Type 100, then press **Tab**. Type 100 a second time, then press **Tab** again. Type 8.

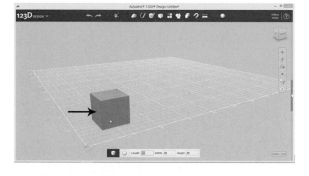

Press **Enter**, then move the cursor to **Top** on the view cube.

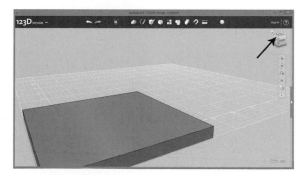

Left-click and release. Go to **Sketch** and drop down to **Ellipse**.

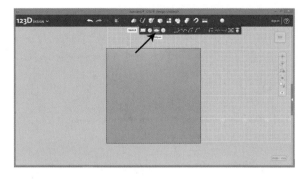

Left-click and release. Move the cursor inside the square.

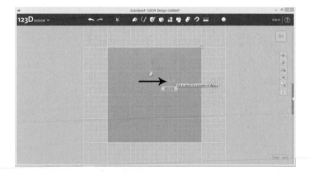

Left-click and release. Move the cursor 65 mm right and 65 mm up from the lower left corner (13 grid marks right, 13 grid marks up). Left-click, release, and move the cursor horizontally to the right.

Type 12.

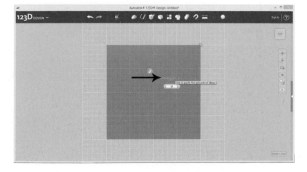

Left-click, release, and pull the cursor downward (straight down from the center point).

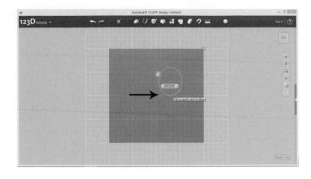

Type 24. Left-click, release, and move the cursor to **Construct**, then drop down to **Extrude**.

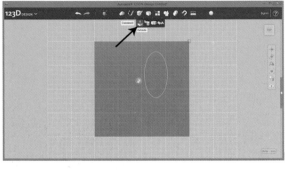

Left-click, release, and bring the cursor inside the ellipse. Left-click and release.

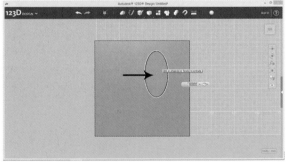

Type −8 to create a hole for the window.

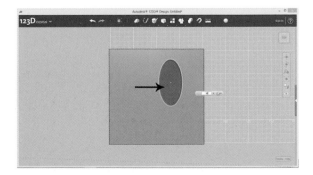

Press **Enter** and move the cursor to **123D** in the upper left corner. On the drop-down menu, select **Export STL** and save your file for printing.

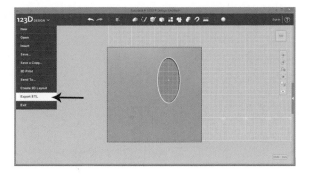

To save the design file, move the cursor to **123D** in the upper left corner and drop down to **Save**.

## Clip

Move the cursor to **Top** on the view cube.

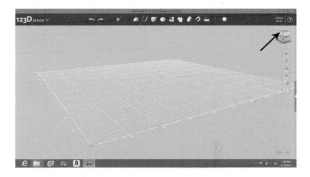

Left-click, release, move the cursor to **Sketch**, and drop down to **Rectangle**.

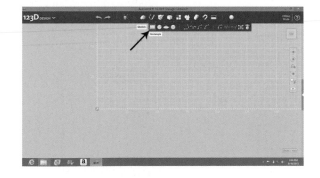

Left-click, release, and move the cursor into the work area. Left-click and release.

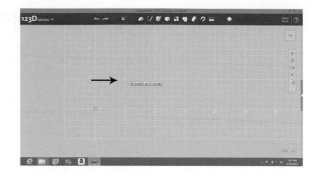

Move the cursor to a grid intersection. Left-click and release. Pull the cursor slightly right and upward.

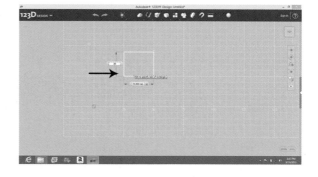

Type 25 (for the horizontal distance). Press the **Tab** key. Type 20.

Left-click and release.

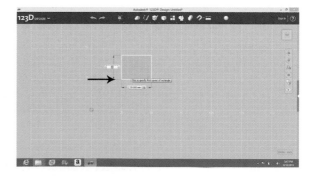

Move the cursor up
1 grid mark from the
lower left corner.

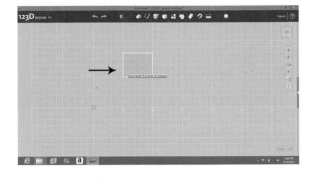

Left-click and release.
Pull the cursor slightly
right and upward.

Type 8 (for the hori-
zontal distance), then
press the **Tab** key. Type
8 . 5, then left-click
and release.

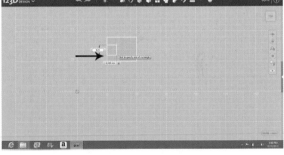

Move the cursor 1 grid
mark left from the
upper right corner.

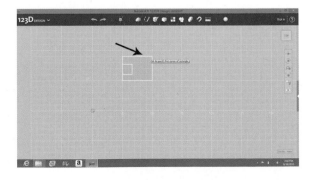

Left-click and release. Type 8 . 5 (for the horizontal distance), then press the **Tab** key. Type 15 (for the vertical distance).

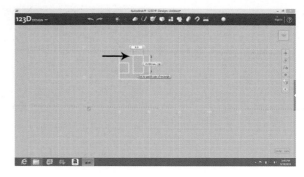

Left-click, release, move the cursor to **Construct**, then drop down to **Extrude**.

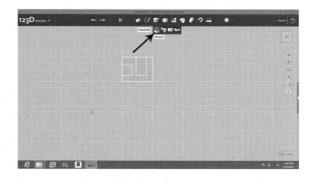

Left-click and release. Move the cursor to the lower portion of the rectangle you created. Left-click and release.

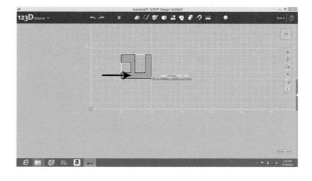

Type 15.

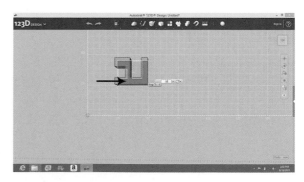

Press **Enter** and move the cursor to **123D** in the upper left corner. On the drop-down menu, choose **Export STL** and save your file for printing.

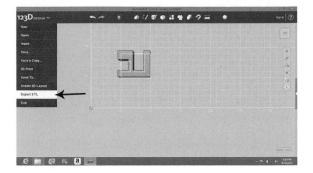

To save the design file, move the cursor to **123D** in the upper left corner and drop down to **Save**.

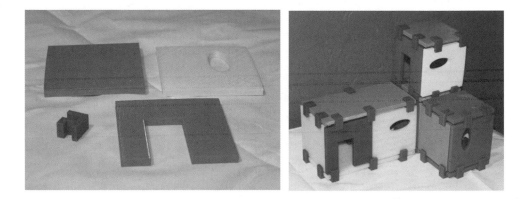

## MAKING THE DOLLHOUSE USING SKETCHUP

### Floor, Wall, or Ceiling Piece

Go to **Camera** and drop down to **Standard Views**. Go to **Top**, then left-click and release.

Go to the **Shape/Rectangle** icon. Left-click and release. Move the dot to the intersection of the red and green lines. Left-click and hold. Pull the cursor slightly right and upward. Type 100,100 and press **Enter**. Release the left-click.

Go to the **Zoom** icon. Left-click and release. Bring the cursor to the small square. Scroll until the square fills about one-third of the screen.

Go to the **Push/Pull** icon. Left-click and release. Bring the cursor inside the square. Left-click, hold, and move the cursor upward slightly. Type 8, then press the **Enter** key. Release the left-click.

Move the cursor to the **Select** icon. Left-click and release. Go to **Edit**, then drop down to **Select All**. Left-click and release.

Go to the menu at the top and choose **File**, then drop down to **Export STL**. Left-click. This file can be used to print the floor, wall, or ceiling pieces.

To save the design file, choose **File**, then **Save**.

## Door

Go to **Camera** and drop down to **Standard Views**. Go to **Top**, then left-click and release.

Go to the **Shape/Rectangle** icon. Left-click and release. Move the dot to the intersection of the red and green lines. Left-click and hold. Pull the cursor slightly right and upward. Type 100,100 and press **Enter**. Release the left-click.

Go to the **Zoom** icon. Left-click and release. Bring the cursor to the small square. Scroll until the square fills about one-third of the screen.

Go to the **Push/Pull** icon. Left-click and release. Bring the cursor inside the square. Left-click, hold, and move the cursor upward slightly. Type 8, then press the **Enter** key. Release the left-click.

Go to the **Tape Measure** icon. Left-click and release. Move the icon to the lower left corner of the square. Left-click, hold, and move the cursor horizontally right. Type 40, then press **Enter**. Release the left-click.

Go to the **Shape/Rectangle** icon. Left-click and release.

Bring the cursor point to the guide point you created with the tape measure. Left-click and hold. Move the cursor up a good bit and right a tiny bit. (If you move the cursor more to the right than up, the coordinates "65,35" will cause the door to be sideways.) Type 65,35 and press **Enter**. Release the left-click.

Go to the **Push/Pull** icon, then left-click and release. Bring the cursor inside the door. Left-click, hold, and pull the cursor upward slightly. Type 8, then press **Enter**. Release the left-click.

Move the cursor to the **Select** icon. Left-click and release. Go to **Edit**, then drop down to **Select All**. Left-click and release.

Go to the menu at the top and choose **File**, then drop down to **Export STL**. Left-click. This file can be used to print a door.

To save the design file, choose **File**, then **Save**.

## Window

Go to **Camera** and drop down to **Standard Views**. Go to **Top**, then left-click and release.

Go to the **Shape/Rectangle** icon. Left-click and release. Move the dot to the intersection of the red and green lines. Left-click and hold. Pull the cursor slightly right and upward. Type 100,100 and press **Enter**. Release the left-click.

Go to the **Zoom** icon. Left-click and release. Bring the cursor to the small square. Scroll until the square fills about one-third of the screen.

Go to the **Push/Pull** icon. Left-click and release. Bring the cursor inside the square. Left-click, hold, and move the cursor upward slightly. Type 8, then press the **Enter** key. Release the left-click.

Go to the **Tape Measure** icon. Left-click and release. Move the point to the lower left corner of the square. Left-click, hold, and move the cursor horizontally right. Type 65, then press **Enter**. Release the left-click.

Move the point to the guide point you just created. Left-click, hold, and pull the cursor straight upward. Type 77, then press **Enter**. Release the left-click.

Go to the **Shape/Circle** icon. Left-click and release. Pull the icon's center to the guide point you just created.

Left-click, hold, and move the point away from the center. Type 12, then press **Enter**. Release the left-click.

Go to the **Scale** icon. Left-click and release. Move the cursor inside the circle. (The circle will fill with small black dots.) Left-click and release.

Move the cursor toward the nine o'clock position until the two center squares turn red. Left-click, hold, and pull the cursor downward. Type 2, then press **Enter**. Release the left-click.

Go to the **Push/Pull** icon. Left-click and release.

Bring the cursor inside the oval. Left-click, hold, and pull the cursor down slightly. Type 8, then press **Enter**. Release the left-click.

Move the cursor to the **Select** icon. Left-click and release. Go to **Edit**, then drop down to **Select All**. Left-click and release.

Go to the menu at the top and choose **File**, then drop down to **Export STL**. Left-click. This file can be used to print a window.

To save the design file, choose **File**, then **Save**.

## Clip

Go to **Camera** and drop down to **Standard Views**. Go to **Top**, then left-click and release.

Go to the **Shape/Rectangle** icon. Left-click and release. Move the dot to the intersection of the red and green lines. Left-click and hold. Pull the cursor to the right and slightly upward. Type 25,20 and press **Enter**. Release the left-click.

Go to the **Zoom** icon. Left-click and release. Bring the cursor to the small square. Scroll until the square fills about one-third of the screen.

Go to the **Tape Measure** icon. Left-click and release. Bring the point to the lower left corner of the rectangle.

Left-click, hold, and pull the cursor straight up. Type 5, then press **Enter**. Release the left-click.

Go to the **Shape/Rectangle** icon. Left-click and release. Bring the cursor point to the guide point you just created. Left-click, hold, and pull the cursor to the right and slightly upward. Type 8,8.5, then press **Enter**. Release the left-click.

Go to the **Tape Measure** icon. Left-click and release. Bring the pointer to the upper right corner of the rectangle. Left-click, hold, and move the cursor horizontally left. Type 5, then press **Enter**. Release the left-click.

Go to the **Shape/Rectangle** icon. Left-click and release. Move the cursor point to the guide point just created. Left-click, hold, and move the cursor to the left and slightly downward. (If you move the cursor more downward than left, the software will expect the first number typed to be a downward number). Type 8.5,15, then press **Enter**. Release the left-click.

Move the cursor to the **Select** icon. Left-click and release. Move the pointer inside the rectangle you just created. Left-click and release. Go to **Edit** and drop down to **Delete**. Left-click and release.

Move the cursor inside the small square you created. Left-click and release. Go to **Edit**, the drop down to **Delete**. Left-click and release.

Go to the **Push/Pull** icon. Left-click and release. Bring the cursor inside the blue figure. Left-click, hold, and pull the cursor downward. Type 8, then press **Enter**. Release the left-click.

Move the cursor to the **Select** icon. Left-click and release. Go to **Edit**, then drop down to **Select All**. Left-click and release.

Go to the menu at the top and choose **File**, then drop down to **Export STL**. Left-click. This file can be used to print the clips.

To save the design file, choose **File**, then **Save**.

# 8

# Catapult

Now you will design a catapult made from two pieces: a base and an arm. A rubber band provides the power for launching.

## Catapult Base

Start a new project. Go to **Primitives** and drop down to **Box**.

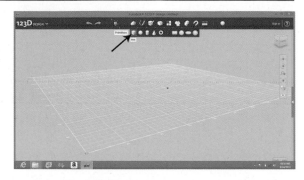

Left-click, release, and move the box to the lower portion of the work area.

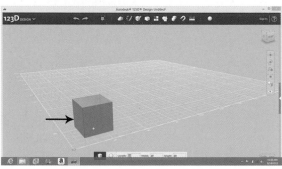

Type 40, then press **Tab**. Type 100, then press **Tab**. Type 45.

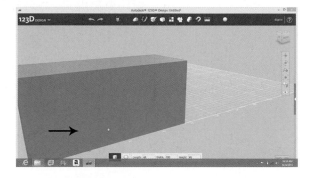

Press **Enter**, then move the cursor to **Front** on the view cube.

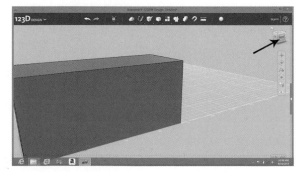

Left-click, release, move the cursor to **Sketch**, and drop down to **Rectangle**.

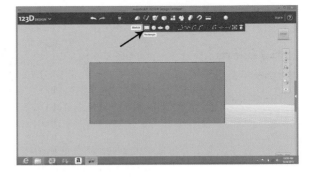

Left-click and release. Move the cursor inside the rectangle.

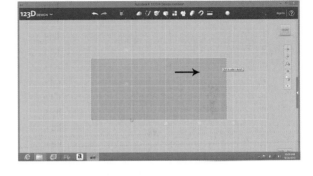

Left-click, release, and move the cursor to the upper right corner of the rectangle.

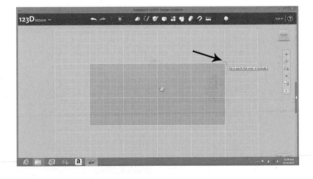

Left-click, release, and pull the cursor down and slightly left. Type 20 (for the horizontal distance), then press the **Tab** key. Type 35 for the vertical distance.

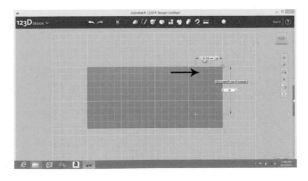

Left-click, release, and move the cursor to the fifth grid mark left of the upper right corner of the rectangle.

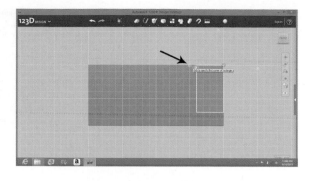

Left-click and release. Pull the cursor down and slightly left. Type 12 for the horizontal distance, then press the **Tab** key. Type 15.

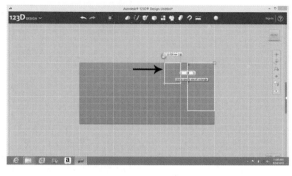

Left-click and release. Move the cursor to the upper left corner.

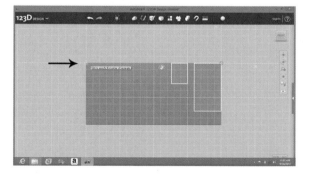

Left-click and release. Pull the cursor down and slightly right. Type 55, then press **Tab**. Type 35.

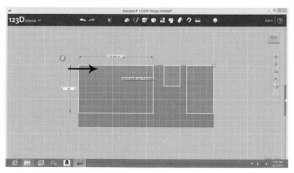

Left-click and release. Move the cursor to **Construct**, then drop down to **Extrude**.

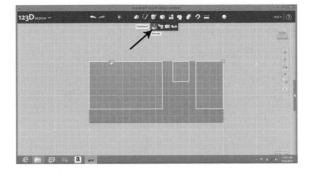

Left-click and release. Move the cursor into the last rectangle you drew. Left-click and release.

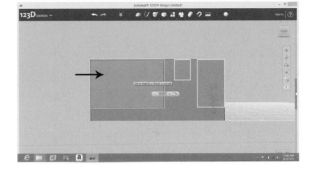

Move the cursor to the rectangle in the center. Left-click and release.

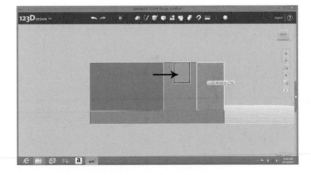

Move the cursor into the rectangle at the right. Left-click and release.

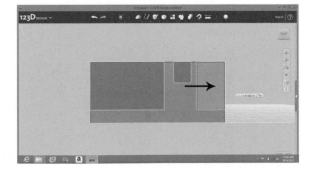

Move the cursor inside the little "0.00mm" box. Left-click and release.

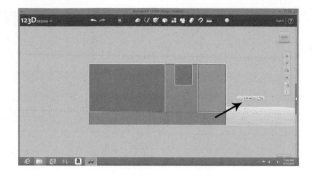

**Backspace** until the box has no numbers or letters. Type −40.

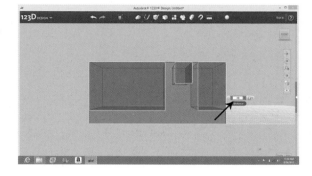

Press **Enter**. Move the cursor to the **Right** side of the view cube.

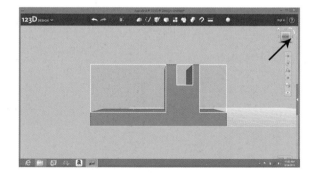

Left-click and release. Move the cursor to **Sketch** and drop down to **Rectangle**.

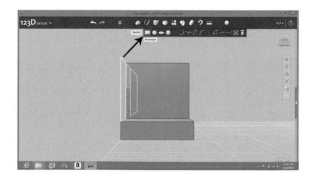

Left-click and release. Move the cursor into the upper rectangle.

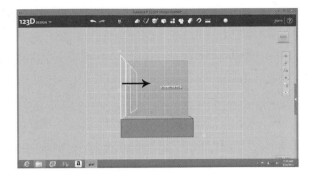

Left-click and release. Move the cursor to a point 2 grid marks left from the upper right corner. Left-click and release. Pull the cursor down and slightly left.

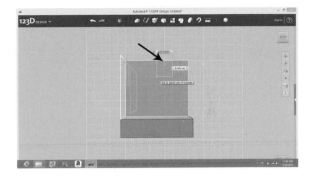

Type 22, then press **Tab**. Type 35.

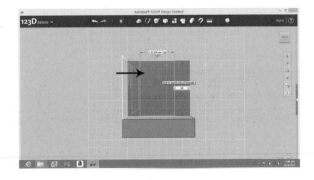

Left-click and release. Move the cursor to **Construct** and drop down to **Extrude**.

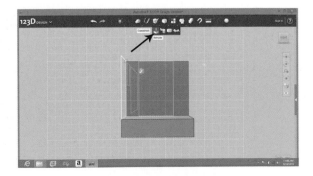

Left-click and release. Move the cursor inside the rectangle you just defined. Left-click and release.

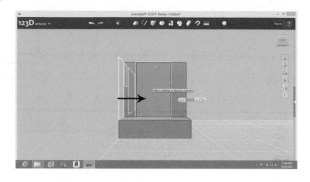

Type −25.

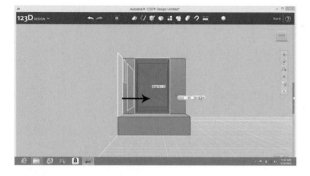

Press **Enter** and move the cursor to **123D** in the upper left corner. On the drop-down menu, select **Export STL** and save your file for printing.

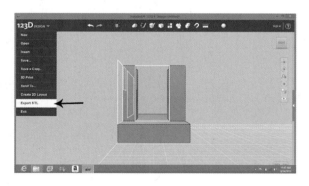

To save the design file, move the cursor to **123D** in the upper left corner and drop down to **Save**.

# Catapult Arm

Start a new project. Go to **Primitives** and drop down to **Box**. Left-click and release.

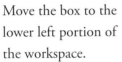

Move the box to the lower left portion of the workspace.

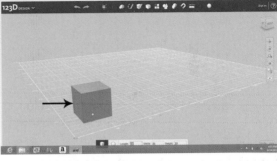

Type 20, then press the **Tab** key. Type 120, then press the **Tab** key. Type 10.

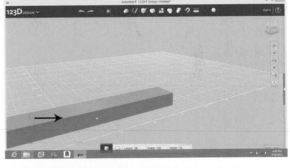

Press **Enter**, then move the cursor to **Front** on the view cube.

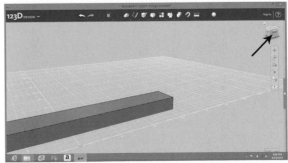

Left-click and release. Move the cursor to **Sketch** and drop down to **Circle**.

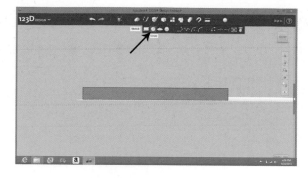

Left-click and release. Move the cursor inside the rectangle. Left-click and release.

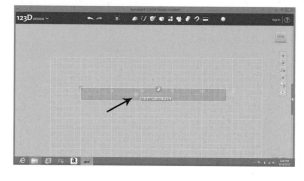

Move the cursor 45 mm right and 5 mm up from the lower left corner (9 grid marks right, 1 grid mark up). Left-click and release. Move the cursor slightly.

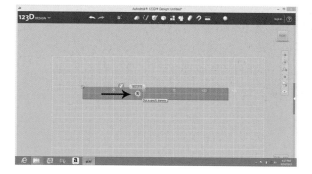

Type 10. Left-click and release.

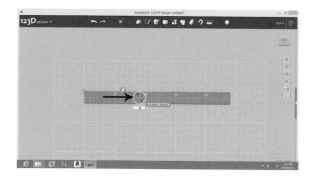

Move the cursor to **Construct** and drop down to **Extrude**. Left-click and release.

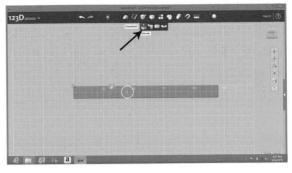

Move the cursor inside the circle. Left-click and release. Type −30. This will cause the action to occur in a downward direction.

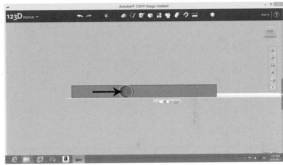

Go to the small black arrow on the far right of the box where you just typed "-30." Left-click on this arrow, then release.

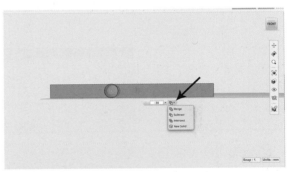

Drop down to **Merge**. Left-click and release. Although the software usually "guesses" what you want to do, it's not always correct. This time, instead of cutting a hole, you want to

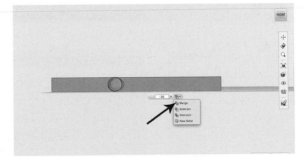

build a round axle on the back side of the rectangle, so you chose **Merge** instead of **Subtract**.

Press **Enter**. Move the cursor to **Construct** and drop down to **Extrude**.

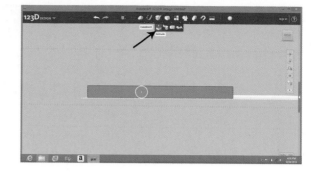

Left-click, release, and move the cursor inside the circle. Left-click and release.

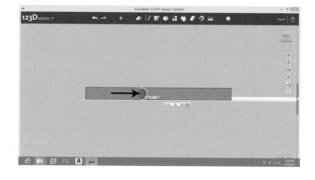

Type 10.

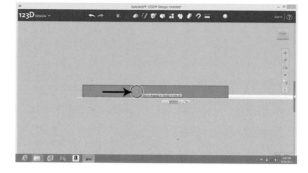

Press **Enter**. Move the cursor to **Top** on the view cube.

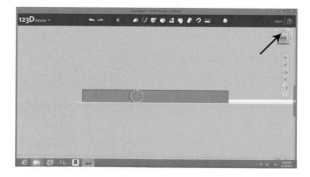

Left-click and release. Move the cursor to **Sketch** and drop down to **Circle**.

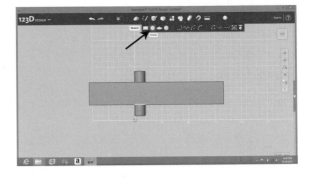

Left-click and release. Move the cursor inside the rectangle. Left-click and release.

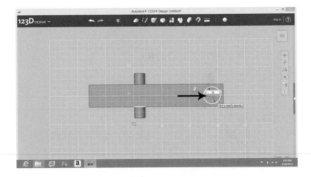

Starting at the lower right corner, move the cursor left 2 grid marks and up 2 grid marks. Left-click and release. Move the cursor slightly. Type 15.

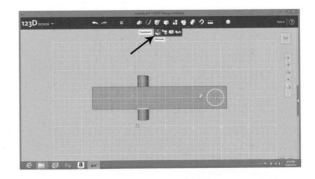

Left-click and release. Move the cursor to **Construct** and drop down to **Extrude**.

Left-click and release. Move the cursor to the inside of the circle. Left-click and release.

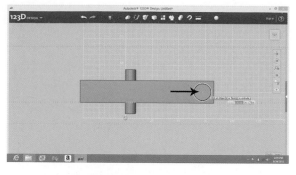

Type – 5 to create a recess.

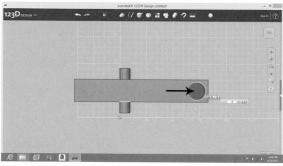

Press **Enter** and move the cursor to **123D** in the upper left corner. On the drop-down menu, select **Export STL** and save your file for printing.

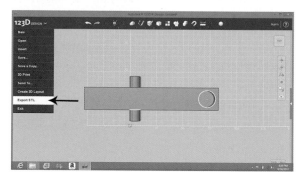

To save the design file, move the cursor to **123D** in the upper left corner and drop down to **Save**.

# MAKING THE CATAPULT USING SKETCHUP

## Catapult Base

Open a new project. Go to **Camera** and drop down to **Standard Views**. Choose **Front**. Left-click and release.

Go to the **Shape/Rectangle** icon. Left-click and release. Move the cursor dot to the intersection of the red and blue lines. Left-click and hold. Move the cursor right and slightly upward. Type 100,45. Press **Enter**. Release the left-click.

Go to the **Zoom** icon. Left-click and release. Move the cursor to the rectangle. Scroll until the rectangle fills half the screen.

Go to the **Shape/Rectangle** icon. Left-click and release. Move the dot to the upper left corner of the rectangle you just created.

Left-click and hold. Move the cursor right and slightly downward. Type 55,35, then press **Enter**. Release the left-click.

Move the cursor dot to the upper right corner. Left-click and hold. Move the cursor left and slightly downward. Type 20,35, then press **Enter**. Release the left-click.

Go to the **Select** icon. Left-click and release. Move the pointer into upper left rectangle. Left-click and release. Go to **Edit** and drop down to **Delete**. Left-click and release. Move the pointer into the upper right rectangle. Left-click and release. Go to **Edit** and drop down to **Delete**. Left-click and release.

Go to the **Push/Pull** icon. Left-click and release. Bring the cursor into the white area. Left-click and hold. Move the cursor upward. Type 40, then press **Enter**. Release the left-click.

Go to **Camera** and drop down to **Standard Views**. Choose **Right**, then left-click and release.

Go to the **Tape Measure** icon. Left-click and release. Move the pointer to the upper right corner of the upper rectangle. Left-click and hold. Move the cursor horizontally left. Type 9, then press **Enter**. Release the left-click.

Go to the **Shape/Rectangle** icon. Left-click and release. Move the pointer until the dot is on top of the guide point you just created.

Left-click and hold. Move the cursor left and slightly downward. Type 22, 35, then press **Enter**. Release the left-click.

Go to the **Push/Pull** icon. Left-click and release. Move the pointer into the rectangle you just created. Left-click and hold. Pull the cursor downward. Type 25, then press **Enter**. Release the left-click.

Go to the **Select** icon. Left-click and release. Move the pointer inside the rectangle you were pushing. Left-click and release. Go to **Edit**, then drop down to **Delete**. Left-click and release.

Go to **Camera** and drop down to **Standard Views**. Choose **Front**, then left-click and release.

Go to the **Tape Measure** icon. Left-click and release. Move the pointer to the upper right corner of the white figure. Left-click and hold. Move the cursor to the left. Type 5, then press **Enter**. Release the left-click.

Move the cursor to the **Shape/Rectangle** icon, then left-click and release. Move the pointer to the guide point you just created. Left-click and hold. Move the cursor left, then slightly down. Type 12, 15, then press **Enter**. Release the left-click.

Go to the **Push/Pull** icon. Left-click and release. Move the cursor inside the rectangle you just created. Left-click and hold. Pull the cursor downward. Type 9, then press **Enter**. Release the left-click.

Go to **Camera** and drop down to **Standard Views**. Choose **Back**, then left-click and release.

Go to the **Tape Measure** icon. Left-click and release. Move the point to the upper left corner of the white figure.

Left-click and hold. Move the cursor right. Type 5, then press **Enter**. Release the left-click.

Go to the **Shape/Rectangle** icon. Left-click and release. Move the dot to the guide point you just created. Left-click and hold. Move the cursor right and slightly down. Type 12, 15, then press **Enter**. Release the left-click.

Move the cursor to the **Push/Pull** icon. Left-click and release. Move the cursor inside the rectangle you just drew. Left-click and hold. Move the cursor down. Type 9, then press **Enter**. Release the left-click.

Move the cursor to the **Select** icon. Left-click and release. Go to **Edit**, then drop down to **Select All**. Left-click and release.

Go to the menu at the top and choose **File**, then drop down to **Export STL**. Left-click. This file can be used to print the catapult base.

To save the design file, choose **File**, then **Save**.

## Catapult Arm

Start a new project. Go to **Camera** and drop down to **Standard Views**. Choose **Front**, then left-click and release.

Go to the **Shape/Rectangle** icon. Left-click and release. Move the dot to the intersection of the red and blue lines. Left-click and hold. Move the cursor right and upward. Type 120,10, then press **Enter**. Release the left-click.

Go to the **Zoom** icon. Left-click and release. Move the cursor to the rectangle (it will look like a tiny black blob), then scroll until the rectangle fills most of the screen.

Go to the **Push/Pull** icon. Left-click and release. Move the cursor inside the white rectangle. Left-click and hold. Move the cursor upward slightly. Type 20, then press **Enter**. Release the left-click.

Go to the **Tape Measure** icon. Left-click and release. Move the dot to the lower left corner of the rectangle. Left-click and hold. Move the cursor to the right. Type 45, then press **Enter**. Release the left-click.

Move the dot until it lines up with the guide point you just created. Left-click and hold. Move the cursor straight upward. Type 5, then press **Enter**. Release the left-click.

Go to the **Shape/Circle** icon. Left-click and release. Move the dot until it latches onto the last guide point you made. Left-click and hold. Move the cursor away from the dot. Type 5, then press **Enter**. Release the left-click.

Go to the **Push/Pull** icon. Left-click and release. Move the cursor into the circle. Left-click and hold. Move the cursor to the left. Type 10, then press **Enter**. Release the left-click.

Go to **Camera**, then drop down to **Standard Views**. Go to **Back**, then left-click and release.

Go to the **Tape Measure** icon. Left-click and release. Move the dot to the lower right corner of the rectangle. Left-click and hold. Move the cursor to the left. Type 4 5, then press **Enter**. Release the left-click.

Move the dot until it lines up with the guide point you just created. Left-click and hold. Move the cursor straight upward. Type 5, then press **Enter**. Release the left-click.

Go to the **Shape/Circle** icon. Left-click and release. Move the dot until it latches onto the last guide point you made. Left-click and hold. Move the cursor away from the dot. Type 5, then press **Enter**. Release the left-click.

Go to the **Push/Pull** icon. Left-click and release. Move the cursor into the circle. Left-click and hold. Move the cursor to the right. Type 10, then press **Enter**. Release the left-click.

Go to **Camera** and drop down to **Standard Views**. Choose **Top**, then left-click and release.

Go to the **Tape Measure** icon. Left-click and release. Move the dot to the lower right corner of the rectangle. Left-click and hold. Move the cursor to the left. Type 10, then press **Enter**. Release the left-click.

Move the dot until it lines up with the guide point you just created. Left-click and hold. Move the cursor straight upward. Type 10, then press **Enter**. Release the left-click.

Go to the **Shape/Circle** icon. Left-click and release. Move the dot until it lines up with the last guide point you created. Left-click and hold. Move the cursor away from the dot. Type 7 . 5, then press **Enter**. Release the left-click.

Go to the **Push/Pull** icon. Left-click and release. Move the cursor into the circle just created. Left-click and hold. Move the cursor slightly left. Type 5, then press **Enter**. Release the left-click.

Move the cursor to the **Select** icon. Left-click and release. Go to **Edit**, then drop down to **Select All**. Left-click and release.

Go to the menu at the top and choose **File**, then drop down to **Export STL**. Left-click. This file can be used to print the catapult arm.

To save the design file, choose **File**, then **Save**.

# 9

# Train

In this chapter, you will design a train. It will have a body, wheels, and hubs to keep the wheels on the body's axles.

## Body

Go to **Primitives** and
drop down to **Box**.

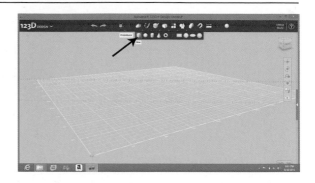

Left-click and release.
Pull the box to the
lower left portion of
the work area. Type
20, then press **Tab**.
Type 80, then press
**Tab**. Type 26.

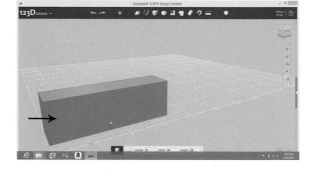

Press **Enter**, then move
the cursor to **Front** on
the view cube.

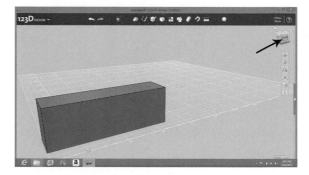

Left-click and release.
Move the cursor to
**Sketch** and drop down
to **Spline**.

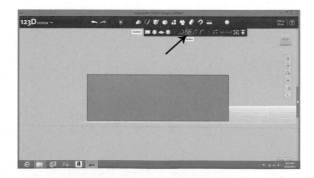

Left-click and release. Move the cursor inside the rectangle. Left-click and release.

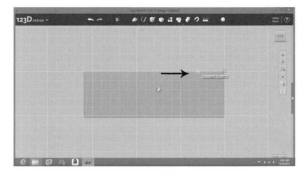

Move the cursor 3 grid marks left from the upper right corner. Left-click and release. Move the cursor down about 2 grid marks and left about 4 grid marks. (It doesn't matter precisely where—you are creating your own shape.)

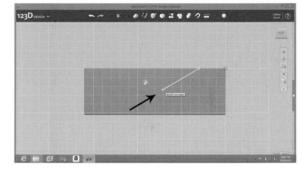

Left-click and release. Move the cursor left and upward.

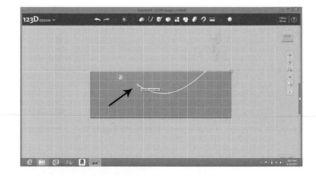

Left-click and release. Move the cursor to about the second grid mark up from the lower left corner.

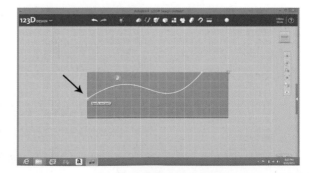

Left-click and release.
Move the cursor left
2 grid marks and up
about 6 grid marks.

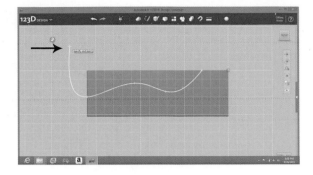

Left-click and release.
Bring the cursor back
to the starting point.
Left-click and release.

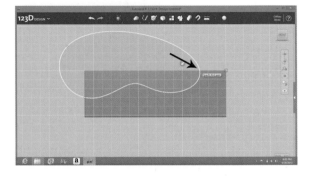

Go to **Construct** and
drop down to **Extrude**.

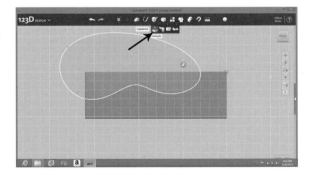

Left-click and release.
Bring the cursor inside
the shape you just
made. Left-click and
release.

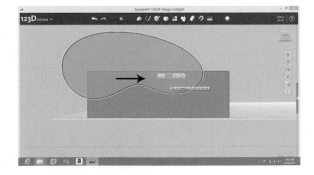

Type -20.

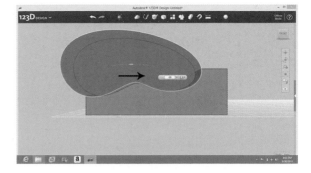

Press **Enter**. Move the
cursor to **Sketch** and
drop down to **Circle**.

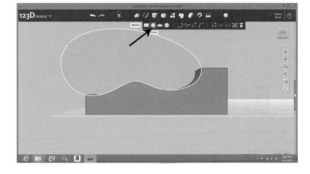

Left-click and release.
Move the cursor to the
lower part of the train
body.

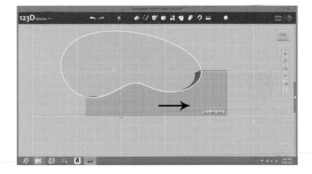

Left-click and release.
Move the cursor left
3 grid marks and up
1 grid mark from the
lower right corner.
Left-click and release.
Move the cursor
slightly.

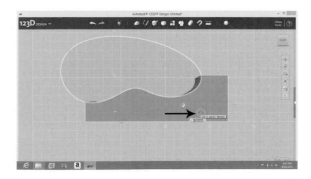

Type 4 . 5.

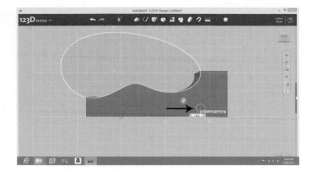

Left-click and release. From the lower left corner, move the cursor 3 grid marks right and 1 grid mark up. Left-click and release. Move the cursor slightly. Type 4 . 5.

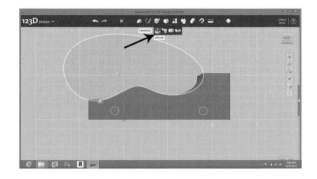

Left-click and release. Move the cursor to **Construct** and drop down to **Extrude**.

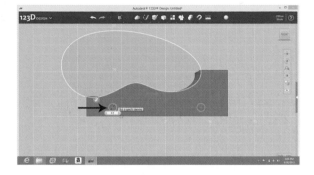

Left-click and release. Move the cursor inside the left circle. Left-click and release.

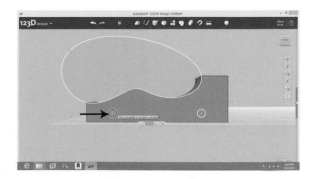

Move the cursor inside the right circle. Left-click and release.

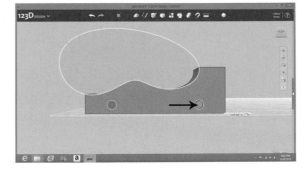

Move the cursor inside the small white box (to the right side). Left-click and release. Press the **Backspace** key until the numbers and letters are gone.

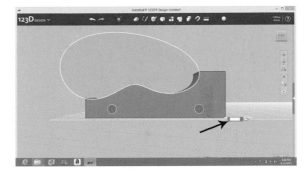

Type -28. This is the distance through the body and 8 mm out the other side. It creates axles at the back of the train.

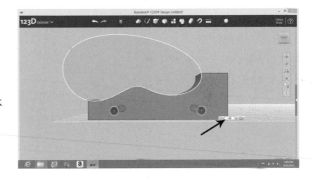

Move the cursor to the right of the white box.

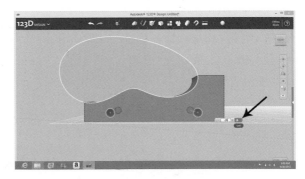

Left-click and release. Drop down to **Merge**.

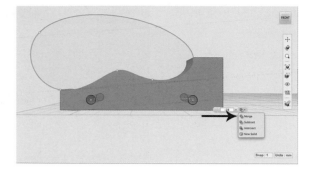

Left-click and release.

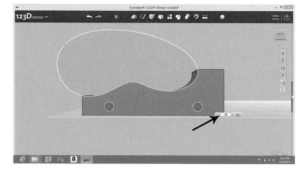

Press **Enter**. Move the cursor to **Construct** and drop down to **Extrude**.

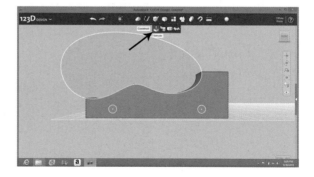

Left-click and release. Move the cursor inside the left circle. Left-click and release.

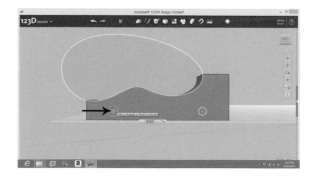

Move the cursor inside the right circle. Left-click and release.

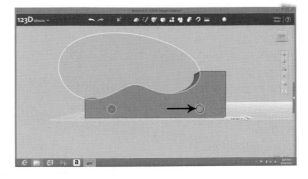

Move the cursor inside the white box (right end of the box). Left-click and release. Press the **Backspace** key until the box is empty. Type 8. This creates the axles at the front of the train.

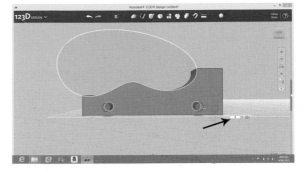

Press **Enter** and move the cursor to **123D** in the upper left corner. On the drop-down menu, choose **Export** STL to save your file for printing.

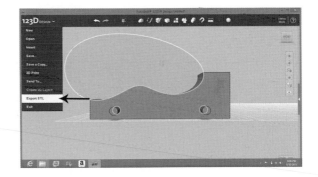

To save the design file, move the cursor to **123D** in the upper left corner and drop down to **Save**.

## Wheels

To design a wheel, you will make a thick circle and cut a hole in the center.

Go to **Primitives** and drop down to **Cylinder**.

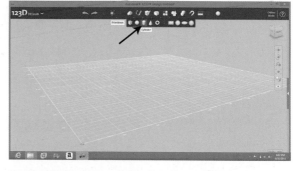

Left-click and release. Pull the cylinder to the lower left in the work area. Type 12, then press the **Tab** key. Type 5.

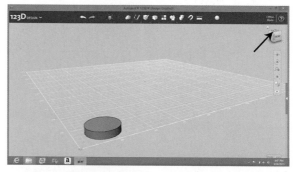

Press **Enter**. Move the cursor to **Top** on the view cube.

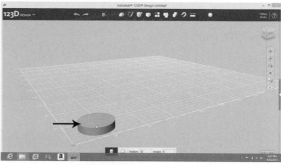

Left-click and release. Move the cursor to **Sketch** and drop down to **Circle**.

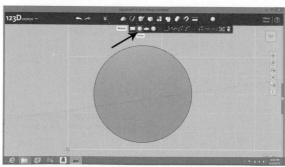

Left-click and release. Move the cursor inside the circle.

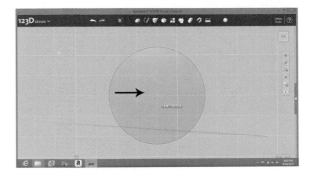

Left-click and release. Move the cursor to the center of the circle (intersection of grid lines). Left-click and release. Move the cursor slightly.

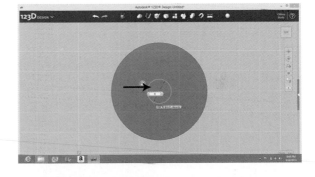

Type 6.

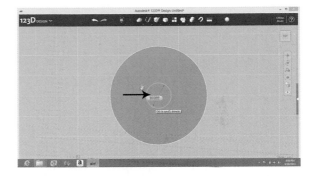

Left-click and release. Move the cursor to **Construct** and drop down to **Extrude**.

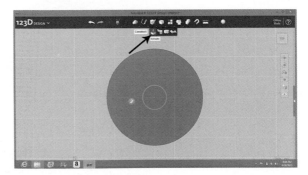

Left-click and release. Move the cursor inside the inner circle. Left-click and release.

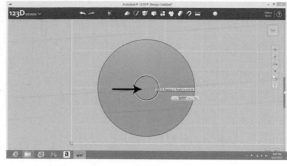

Type −5.

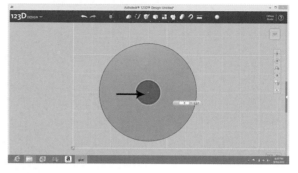

Press **Enter** and move the cursor to **123D** in the upper left corner. On the drop-down menu, select **Export STL** and save your file for printing. You will need to print four wheels.

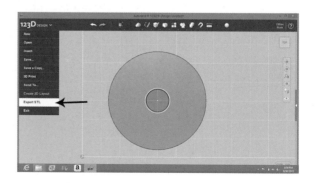

To save the design file, move the cursor to **123D** in the upper left corner and drop down to **Save**.

# Hub

Go to **Primitives**
and drop down to
**Cylinder**.

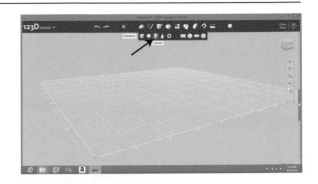

Left-click and release.
Pull the cylinder to the
lower left of the screen.
Type 4 for the radius,
then press **Tab**. Type 4
for the height as well.

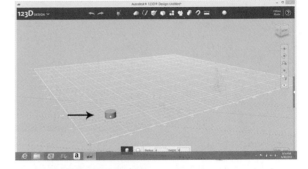

Press **Enter**. Move the
cursor to **Top** on the
view cube.

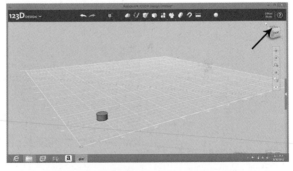

Left-click and release.
Move to **Sketch** and
drop down to **Circle**.

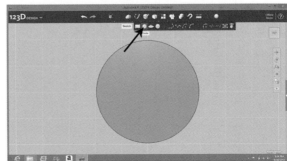

Left-click and release.
Move the cursor inside
the circle. Left-click
and release. Move the
cursor to the center of
the circle. Left-click
and release. Move the
cursor slightly.

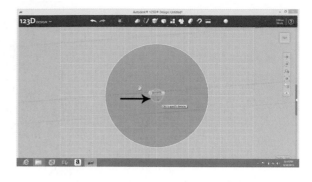

Type 5 for the
diameter.

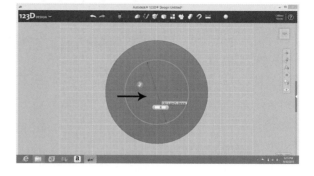

Left-click and release.
Go to **Construct** and
drop down to **Extrude**.

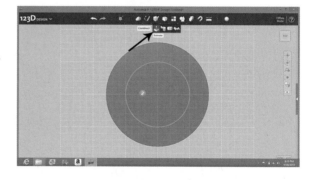

Left-click and release.
Move the cursor inside
the smaller circle. Left-
click and release.

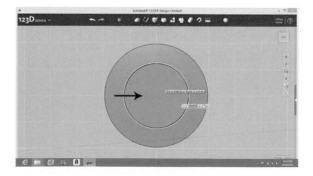

Type –2.

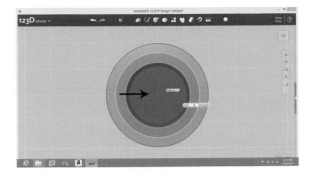

Press **Enter** and move
the cursor to **123D** in
the upper left corner.
On the drop-down
menu, select **Export
STL** and save your file
for printing. You will
need to print four of
these.

To save the design file, move the cursor to **123D** in the upper left
corner and drop down to **Save**.

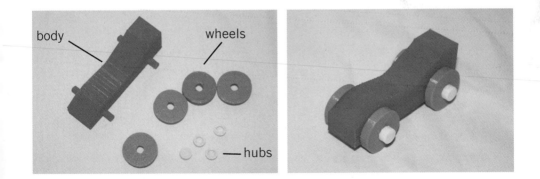

body   wheels   hubs

## MAKING THE TRAIN USING SKETCHUP

## Body

Start a new project. Go to **Camera** and drop down to **Standard Views**. Go to **Front**, then left-click and release.

Select the **Shape/Rectangle** icon. Left-click and release. Move the dot to the origin. Left-click and hold. Move the cursor to the right and slightly upward. Type 80,26 and press **Enter**. Release the left-click.

Go to the **Zoom** icon. Left-click and release. Bring it down to the rectangle, then scroll until the rectangle fills about half the screen.

Go to the **Tape Measure** icon. Left-click and release. Move the pointer to the upper right corner of the rectangle. Left-click and hold. Pull the cursor horizontally to the left. Type 15 and press **Enter**. Release the left-click.

Go to **Lines** (third icon from the left) and drop down to **Freehand**. Left-click and release. Move the dot to the guide point created with the **Tape Measure** tool. Left-click and hold. Draw a shape that looks similar to the one at the bottom of page 141. Release the left-click.

Go to the **Select** icon. Left-click and release. Move the pointer to the top left part of the rectangle. Left-click and release. (Black dots should fill the area.) Go to **Edit** and drop down to **Delete**. Left-click and release.

Go to the **Push/Pull** icon. Left-click and release. Move the pointer into the lower part of the train body. Left-click and hold. Move the pointer. Type 20 and press **Enter**. Release the left-click.

Go to the **Tape Measure** icon. Left-click and release. Move the pointer to the lower left corner. Left-click and hold, then move the cursor to the right. Type 15 and press **Enter**. Release the left-click. Move the pointer to the guide point you just created. Left-click and hold. Pull the pointer straight upward. Type 5, then press **Enter**. Release the left-click.

Move the pointer to the lower right corner. Left-click and hold, then move the cursor to the left. Type 15 and press **Enter**. Release the left-click. Move the pointer to the guide point you just created. Left-click and hold.

Pull the pointer straight upward. Type 5, then press **Enter**. Release the left-click.

Go to the **Shape/Circle** icon. Left-click and release. Move the pointer to the left guide point. Left-click and hold. Move the cursor slightly. Type 2.25 (the radius) and press **Enter**. Release the left-click.

Move the pointer to the right guide point. Left-click and hold. Move the cursor slightly. Type 2.25 (the radius) and press **Enter**. Release the left-click.

Go to the **Push/Pull** icon. Left-click and release. Move the pointer inside the left circle. Left-click and hold. Move the cursor to the left. Type 8, then press **Enter**. Release the left-click.

Move the pointer to the inside of the right circle. Left-click and hold. Move the cursor to the right. Type 8, then press **Enter**. Release the left-click.

Now, you'll do the same thing on the other side. Go to **Camera** and drop down to **Standard Views**. Choose **Back**, then left-click and release.

Go to the **Tape Measure** icon. Left-click and release. Move the pointer to the lower left corner. Left-click and hold, then move the cursor to the right. Type 15 and press **Enter**. Release the left-click. Move the pointer to the guide point you just created. Left-click and hold. Pull the pointer straight upward. Type 5, then press **Enter**. Release the left-click.

Move the pointer to the lower right corner. Left-click and hold, then move the cursor to the left. Type 15 and press **Enter**. Release the left-click. Move the pointer to the guide point you just created. Left-click and hold. Pull the pointer straight upward. Type 5, then press **Enter**. Release the left-click.

Go to the **Shape/Circle** icon. Left-click and release. Move the pointer to the left guide point. Left-click and hold. Move the cursor slightly. Type 2.25 (the radius) and press **Enter**. Release the left-click.

Move the pointer to the right guide point. Left-click and hold. Move the cursor slightly. Type 2.25 (the radius) and press **Enter**. Release the left-click.

Go to the **Push/Pull** icon. Left-click and release. Move the pointer inside the left circle. Left-click and hold. Move the cursor to the left. Type 8, then press **Enter**. Release the left-click.

Move the pointer to the inside of the right circle. Left-click and hold. Move the cursor to the right. Type 8, then press Enter. Release the left-click.

Move the cursor to the **Select** icon. Left-click and release. Go to **Edit**, then drop down to **Select All**. Left-click and release.

Go to the menu at the top and choose **File**, then drop down to **Export STL**. Left-click. This file can be used to print the train body.

To save the design file, choose **File**, then **Save**.

## Wheel

Start a new project. Go to **Camera**, then drop down to **Standard Views**, then select **Front**. Left-click and release.

Go to the **Shape/Circle** icon. Left-click and release.

Bring the pointer dot to the origin. Left-click and hold. Move the cursor slightly. Type 12, then press **Enter**. Release the left-click.

Go to the **Zoom** icon. Left-click and release. Move the icon down and scroll until the circle fills about one-third of the screen.

Go to the **Shape/Circle** icon. Left-click and release. Move the pointer to the origin. Left-click and hold. Move the cursor slightly. Type 3, then press **Enter**. Release the left-click.

Go to the **Select** icon. Left-click and release. Move the pointer to the small inner circle. Left-click and release.

Go to **Edit**, then drop down to **Delete**. Left-click and release.

Go to the **Push/Pull** icon. Left-click and release. Bring the pointer into the blue wheel area. Left-click, hold, and move the pointer. Type 5 and press **Enter**. Release the left-click.

Move the cursor to the **Select** icon. Left-click and release. Go to **Edit**, then drop down to **Select All**. Left-click and release.

Go to the menu at the top and choose **File**, then drop down to **Export STL**. Left-click. This file can be used to print a train wheel. You will need four wheels.

To save the design file, choose **File**, then **Save**.

# Hub

Start a new project. Go to **Camera** and drop down to **Standard Views**, then select **Front**. Left-click and release.

Go to the **Shape/Circle** icon. Left-click and release.

Bring the pointer dot to the origin. Left-click and hold. Move the cursor slightly. Type 4, then press **Enter**. Release the left-click.

Go to the **Zoom** icon. Left-click and release. Move the icon down and scroll until the circle fills about one-third of the screen.

Go to the **Push/Pull** icon. Left-click and release. Move the pointer into the lower half of the circle. Left-click and hold. Move the pointer slightly upward. Type 4, then press **Enter**. Release the left-click.

Go to the **Shape/Circle** icon. Left-click and release. Move the pointer to the origin. Left-click and hold. Move the cursor slightly. Type 2.5, then press **Enter**. Release the left-click.

Go to the **Push/Pull** icon. Left-click and release. Move the pointer into the lower half of the small circle. Left-click and hold. Move the pointer slightly upward. Type 2, then press **Enter**. Release the left-click.

Move the cursor to the **Select** icon. Left-click and release. Go to **Edit**, then drop down to **Select All**. Left-click and release.

Go to the menu at the top and choose **File**, then drop down to **Export STL**. Left-click. This file can be used to print the hub. You will need to print four hubs.

To save the design file, choose **File**, then **Save**.

# 10

# Train Track

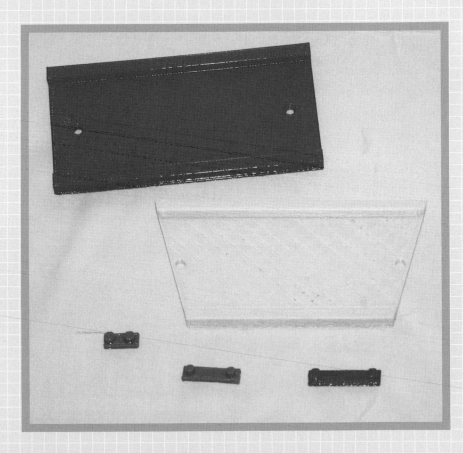

In this chapter, you will design a track for your train. You will create a straight piece, an angled piece to make the track curve, and connectors.

## Straight Piece

Go to **Primitives** and drop down to **Box**.

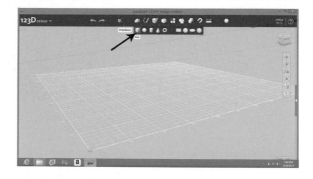

Left-click and release. Move the box to the lower left part of the work area.

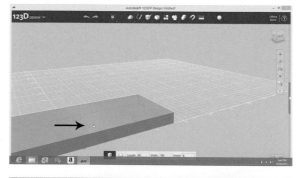

Type 50, then press **Tab**. Type 100, then press **Tab**. Type 8.

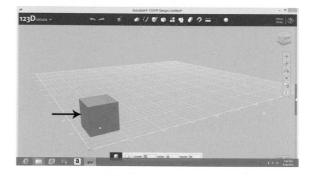

Press **Enter**, then move the cursor to **Top** on the view cube.

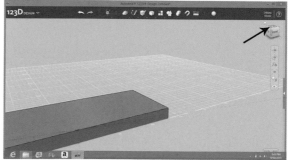

Left-click and release. Move the cursor to **Sketch**, then drop down to **Rectangle**.

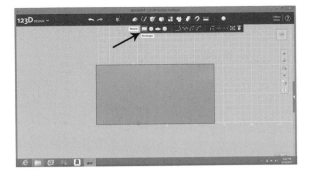

Left-click and release. Move the cursor inside the rectangle.

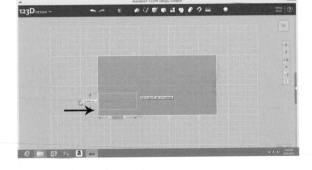

Left-click and release. Move the cursor to the lower left corner, then up 1 grid mark. Left-click and release. Move the cursor right and slightly upward.

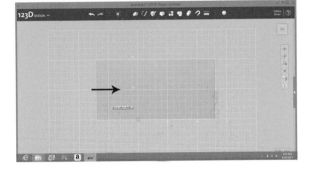

Type 100, then press **Tab**. Type 40.

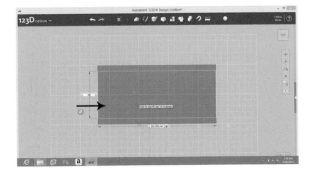

Left-click and release.
Go to **Construct** and
drop down to **Extrude**.

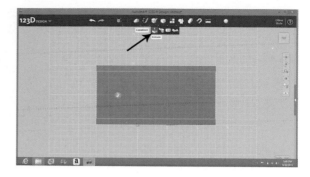

Left-click and release.
Move the cursor into
the center rectangle.
Left-click and release.

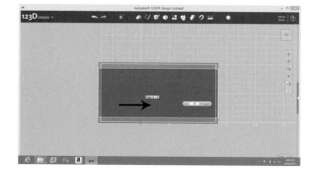

Type −5.

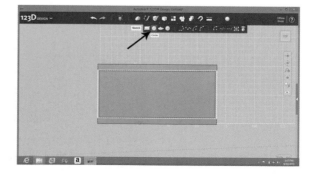

Press **Enter**. Move the
cursor to **Sketch** and
drop down to **Circle**.

Left-click and release.
Move the cursor inside
the smaller rectangle.

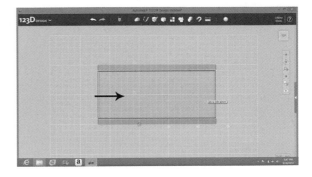

Left-click and release.
Move the cursor to
the lower right corner,
then up 5 grid marks
and left 2 grid marks.
Left-click and release.
Move the cursor
slightly.

Type 4.

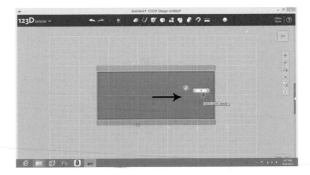

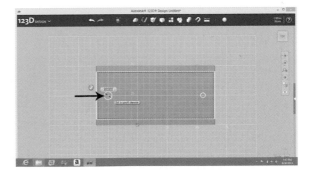

Left-click and release.
Move the cursor to
the lower left corner,
then up 5 grid marks
and right 2 grid marks.
Left-click and release.
Move the cursor
slightly.

Type 4.

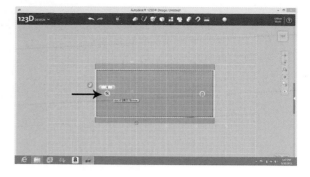

Left-click and release.
Move the cursor to
**Construct**, then drop
down to **Extrude**.

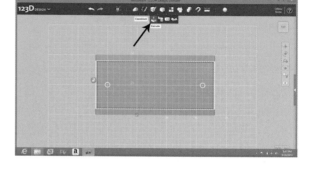

Left-click and release.
Move the cursor inside
the left circle. Left-
click and release.

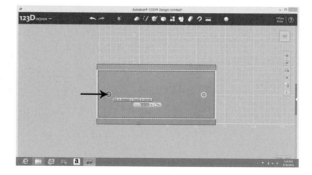

Move the cursor inside
the right circle. Left-
click and release.

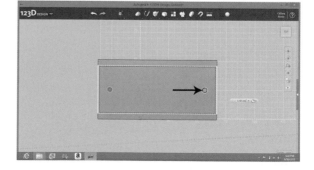

Move the cursor inside the small white box (to the right). Left-click and release. Press **Backspace** until the numbers and letters are gone.

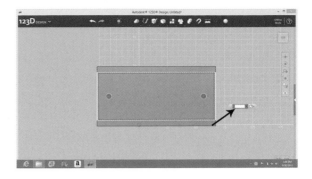

Type -8. Move the cursor right to the small black arrow. Left-click and release. Drop down to **Subtract**, then left-click and release again.

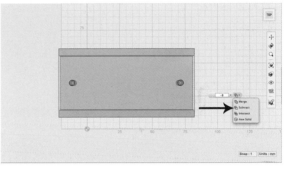

Press **Enter** and move the cursor to **123D** in the upper left corner. On the drop-down menu, choose **Export STL** and save your file for printing.

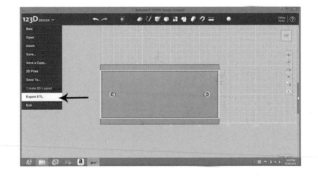

To save the design file, move the cursor to **123D** in the upper left corner and drop down to **Save**.

# Curve

To create the curved section of track, you will start with a straight piece. Either continue from the last step of the straight track, or (if you closed out and went away) load the design file for the straight track.

Go to **Sketch** and drop down to **Polyline**.

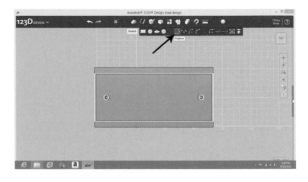

Left-click and release. Move the cursor inside the rectangle, near the lower left corner. Left-click and release.

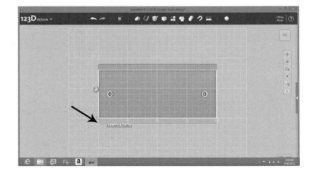

Place the cursor on the lower left corner. Left-click and release. Move the cursor above the rectangle and to the right.

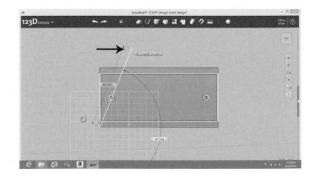

Press the **Tab** key. The little box with "deg" should turn blue. Type 165.

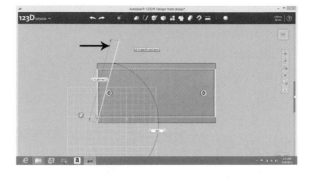

Left-click and release. Move the cursor to the left of the rectangle. It doesn't matter exactly where the line goes—as long as the line doesn't touch the rectangle. Left-click and release.

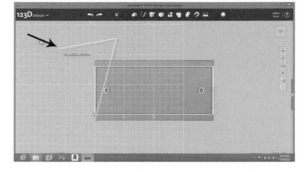

Bring the cursor back to the lower left corner of the rectangle. Left-click and release.

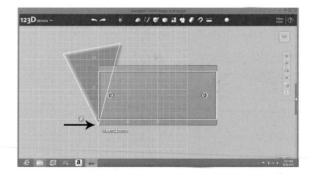

Move the cursor to **Construct** and drop down to **Extrude**.

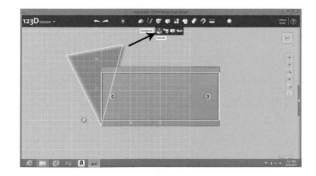

Left-click and release. Move the cursor inside the triangle you just created. Left-click and release.

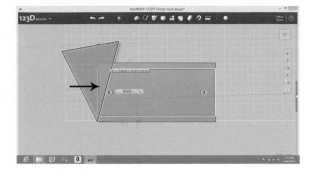

Type −8 to remove the material selected.

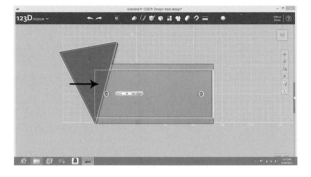

Press **Enter**. Move the cursor to **Sketch** and drop down to **Polyline**.

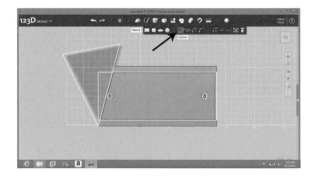

Left-click and release. Move the cursor inside the rectangle, near the lower right corner. Left-click and release.

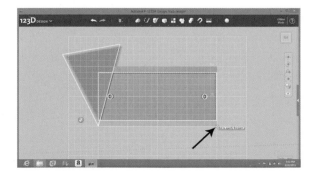

Place the cursor on the lower right corner of the rectangle. Left-click and release. Move the cursor above the rectangle and slightly left.

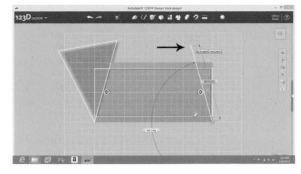

Press the **Tab** key and the box with "deg" will turn blue. Type 165.

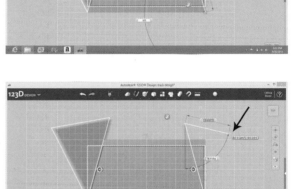

Left-click and release. Move the cursor to the right of the rectangle. Left-click and release.

Move the cursor to the lower right corner of the rectangle. Left-click and release.

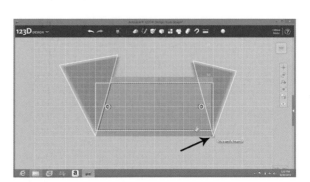

Move the cursor to **Construct** and drop down to **Extrude**.

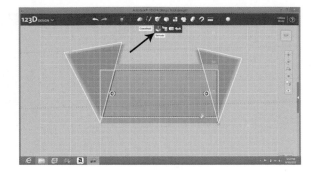

Left-click and release. Move the cursor inside the right triangle. Left-click and release.

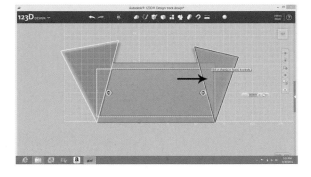

Type −8.

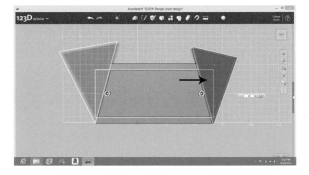

Press **Enter** and move the cursor to **123D** in the upper left corner. On the drop-down menu, select **Export** STL and save your file for printing.

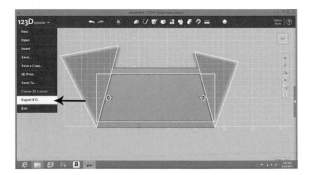

To save the design file for the curve, move the cursor to **123D** in the upper left corner and drop down to **Save a Copy**. If you choose **Save** instead of **Save a Copy**, the operation

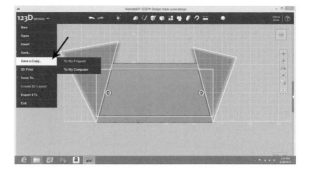

will overwrite the file for a straight piece of track. (Remember that you started this design with the straight piece of track.)

## Curve Connector

Go to **Primitives** and drop down to **Box**.

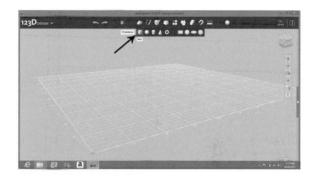

Left-click and release. Move the box to the lower left portion of the work area.

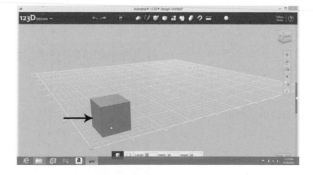

Type 5, then press **Tab**. Type 12, then press **Tab**. Type 2.

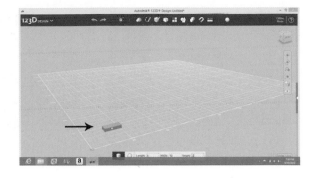

Press **Enter**. Move the cursor to **Top** on the view cube.

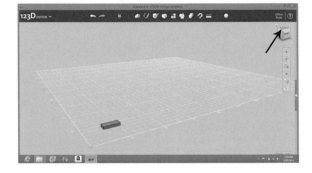

Left-click and release. Move the cursor to **Sketch** and drop down to **Circle**.

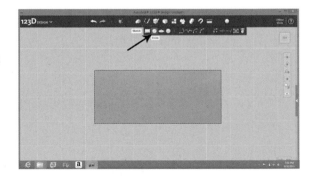

Left-click and release. Move the cursor inside the rectangle. Left-click and release.

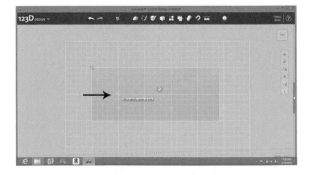

Move the cursor
2.5 mm right and
2.5 mm up from the
lower left corner. Left-
click and release. Move
the cursor slightly.

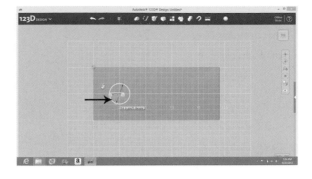

Type 3.5.

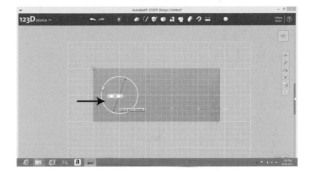

Left-click and release.
Move the cursor
2.5 mm up and
2.5 mm left from the
lower right corner of
the rectangle.

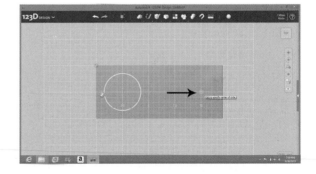

Left-click and release.
Move the cursor
slightly. Type 3.5.

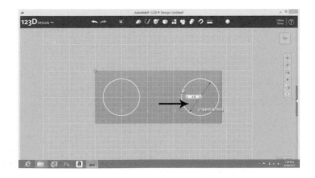

Left-click and release. Move the cursor to **Construct** and drop down to **Extrude**.

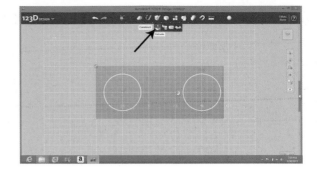

Left-click and release. Move the cursor inside the left circle. Left-click and release.

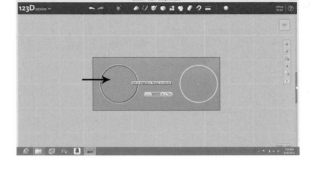

Move the cursor inside the right circle. Left-click and release.

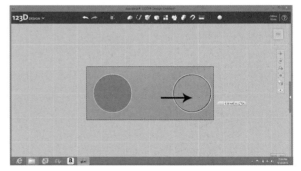

Move the cursor inside the white box (to the right end). Left-click and release. Press **Backspace** until there are no numbers or letters in the box.

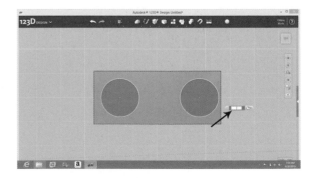

Type 3 to create the
two pegs.

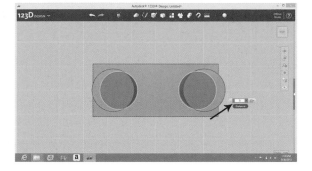

Press **Enter** and move
the cursor to **123D** in
the upper left corner.
On the drop-down
menu, select **Export
STL** and save your file
for printing.

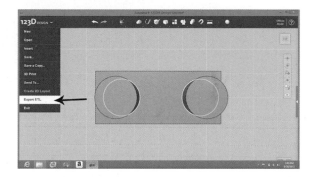

To save the design file, move the cursor to **123D** in the upper left
corner and drop down to **Save**.

## Curve-to-Straight Connector

The process is the same as the curve connector, but the width of the box is
20 and the distance between the center of the circles is 14.

## Straight-to-Straight Connector

The process is the same as the curve connector, but the width of the box is 25 and the distance between the center of the circles is 20.

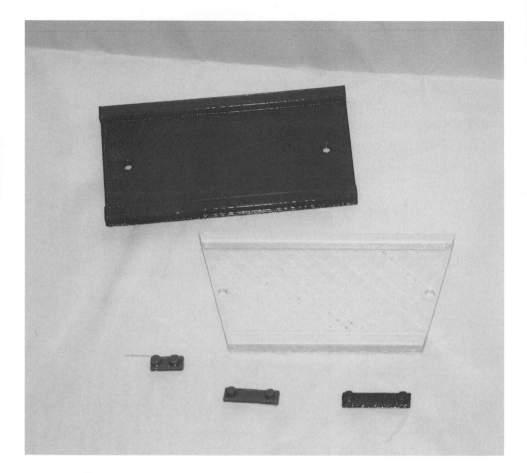

# MAKING TRACK USING SKETCHUP

## Straight Piece

Start with a new file. Go to **Camera**, then drop down to **Standard Views**, then **Front**. Left-click and release.

Choose the **Shape/Rectangle** icon. Left-click and release.

Move the pointer to the intersection of red and blue lines. Left-click and hold. Move the cursor right and slightly up. Type 100,50 and press **Enter**. Release the left-click.

Choose the **Zoom** icon. Left-click and release. Move the pointer to the rectangle. Scroll until the rectangle fills most of the work area.

Choose the **Tape Measure** icon. Left-click and release.

Move the pointer to the lower left corner. Left-click and hold. Move the pointer to the right. Type 10, then press **Enter**. Release the left-click.

Move the pointer to the guide point you just created. Left-click and hold. Move the pointer upward. Type 25, then press **Enter**. Release the left-click.

Move the pointer to the lower right corner. Left-click and hold. Move the pointer to the left. Type 10, then press **Enter**. Release the left-click.

Move the pointer to the guide point you just created. Left-click and hold. Move the pointer upward. Type 25, then press **Enter**. Release the left-click.

Move the pointer to the lower left corner. Left-click and hold. Move the pointer straight up. Type 5, then press **Enter**. Release the left-click.

Move the pointer to the upper left corner. Left-click and hold. Move the pointer straight down. Type 5, then press **Enter**. Release the left-click.

Move the pointer to the upper left corner. Left-click and hold. Move the pointer horizontally to the right. Type 13.4, then press Enter. Release the left-click.

Move the pointer to the upper right corner. Left-click and hold. Move the pointer horizontally to the left. Type 13.4, then press **Enter**. Release the left-click.

Go to **File**, then drop down to **Save As**. Name the file "Template" and save. You will use this file later when you create the curved section of track.

Move the cursor to the **Push/Pull** icon. Left-click and release.

Move the cursor inside the rectangle, toward the bottom. Left-click and hold. Move the cursor upward. Type 8, then press **Enter**. Release the left-click.

Move the cursor to the **Shape/Circle** icon. Left-click and release. Move the pointer to the center left guide point. Left-click and hold. Move the cursor slightly. Type 2, then press **Enter**. Release the left-click.

Move the pointer to the center right guide point. Left-click and hold. Move the cursor slightly. Type 2, then press **Enter**. Release the left-click.

Move the cursor to the **Push/Pull** icon. Left-click and release.

Move the pointer inside the left circle. Left-click and hold. Move the pointer to the right. Type 8, then press **Enter**. Release the left-click.

Move the pointer inside the right circle. Left-click and hold. Move the pointer to the left. Type 8, then press **Enter**. Release the left-click.

Move the cursor to the **Shape/Rectangle** icon. Left-click and release.

Move the pointer to the guide point that is 5 mm above the lower left corner. Left-click and hold. Move the pointer right and slightly upward. Type 100,40 and press **Enter**. Release the left-click.

Move the cursor to the **Push/Pull** icon. Left-click and release.

Move the cursor inside the center rectangle, toward the bottom. Left-click and hold. Pull the cursor upward. Type 5 and press **Enter**. Release the left-click.

Move the cursor to the **Select** icon. Left-click and release. Go to **Edit**, then drop down to **Select All**. Left-click and release.

Go to the menu at the top and choose **File**, then drop down to **Export STL**. Left-click. This file can be used to print straight track.

Before you save your design file, remember that it must be renamed to prevent writing over the template that you saved earlier. To save the design file, choose **File**, then **Save As**. Name the file "Straight Track."

# Curve

Go to **File** and drop down to **Open**. Choose "Template." (This was saved earlier.)

Go to the **Push/Pull** icon. Left-click and release.

Move the pointer inside the rectangle, toward the bottom. Left-click and hold. Bring the pointer upward. Type 8 and press **Enter**. Release the left-click.

Go to **Lines** (third icon from left), then select **Line** from the drop-down menu. Left-click and release.

Move the pointer to the lower left corner. Left-click and hold. Move the pointer to the top of the rectangle, to the guide point 13.4 mm from the top left corner. Release the left-click.

Move the pointer to the lower right corner. Left-click and hold. Move the pointer to the top of the rectangle, to the guide point 13.4 mm from the top right corner. Release the left-click.

Go to the **Push/Pull** icon. Left-click and release.

Move the pointer inside the triangle on the left side of the rectangle. Left-click and hold. Move the cursor slightly to the right. Type 8 and press **Enter**. Release the left-click.

Move the pointer inside the triangle on the right side of the rectangle. Left-click and hold. Move the cursor slightly to the left. Type 8 and press **Enter**. Release the left-click.

Move the cursor to the **Shape/Circle** icon. Left-click and release.

Move the pointer to the left center guide point. Left-click and hold. Move the cursor slightly. Type 2 and press **Enter**. Release the left-click.

Move the pointer to the right center guide point. Left-click and hold. Move the cursor slightly. Type 2 and press **Enter**. Release the left-click.

Move the cursor to **Lines** and drop down to **Line**. Left-click and release.

Move the pointer to the guide point 5 mm up from the lower left corner. Left-click and hold. Move the pointer horizontally to the right until you reach the right edge of the white figure. Release the left-click.

Move the pointer to the guide point 5 mm down from the upper left corner of the original rectangle. Left-click and hold. Move the pointer horizontally to the right until you reach the right edge of the white figure. (The line will be red as you move straight to the right.) Release the left-click.

Move the cursor to the **Push/Pull** icon. Left-click and release.

Move the pointer inside the left circle. Left-click and hold. Move the pointer to the right. Type 8, then press **Enter**. Release the left-click.

Move the pointer inside the right circle. Left-click and hold. Move the pointer to the left. Type 8, then press **Enter**. Release the left-click.

Move the pointer inside the center trapezoid (rectangle with slanted sides), close to the lower edge. Left-click and hold. Move the cursor upward. Type 5 and press **Enter**. Release the left-click.

Move the cursor to the **Select** icon. Left-click and release. Go to **Edit**, then drop down to **Select All**. Left-click and release.

Go to the menu at the top and choose **File**, then drop down to **Export STL**. Left-click. This file can be used to print curved track.

Before you save your design file, remember it must be renamed to prevent writing over the "Template" you saved earlier. To save the design file, choose **File**, then **Save As**. Name the file "Curve."

## Curve Connector

Start with a new file. Go to **Camera** and drop down to **Standard Views**, then **Front**. Left-click and release.

Choose the **Shape/Rectangle** icon. Left-click and release.

Move the pointer to the intersection of red and blue lines. Left-click and hold. Move the cursor right and slightly up. Type 12,5 and press **Enter**. Release the left-click.

Choose the **Zoom** icon. Left-click and release. Move the pointer to the rectangle. Scroll until the rectangle fills most of the work area.

Choose the **Tape Measure** icon. Left-click and release.

Move the pointer to the lower left corner. Left-click and hold. Move the pointer to the right. Type 2.5, then press **Enter**. Release the left-click.

Move the pointer to the guide point you just created. Left-click and hold. Move the pointer up. Type 2.5, then press **Enter**. Release the left-click.

Move the pointer to the lower left corner. Left-click and hold. Move the pointer to the right. Type 10, then press **Enter**. Release the left-click. This is 7.5 mm to the right of the center of the first point

Move the pointer to the guide point you just created. Left-click and hold. Move the pointer up. Type 2.5, then press **Enter**. Release the left-click.

Go to the **Push/Pull** icon. Left-click and release. Move the pointer inside the lower part of the rectangle. Left-click and hold. Move the pointer upward. Type 2, then press **Enter**. Release the left-click.

Move the cursor to the **Shape/Circle** icon. Left-click and release.

Move the pointer to the left guide point. Left-click and hold. Move the pointer slightly. Type 1.75, then press **Enter**. Release the left-click.

Move the pointer to the right guide point. Left-click and hold. Move the pointer slightly. Type 1.75, then press **Enter**. Release the left-click.

Go to the **Push/Pull** icon. Left-click and release. Move the pointer inside the left circle. Left-click and hold. Move the pointer to the left. Type 3, then press **Enter**. Release the left-click.

Move the pointer inside the right circle. Left-click and hold. Move the pointer to the right. Type 3, then press **Enter**. Release the left-click

Move the cursor to the **Select** icon. Left-click and release. Go to **Edit**, then drop down to **Select All**. Left-click and release.

Go to the menu at the top and choose **File**, then drop down to **Export STL**. Left-click. This file can be used to print the curve connector.

To save the design file, choose **File**, then **Save**.


## Curve-to-Straight Connector

The process is the same as the curve connector, but the width of the rectangle is 20 and the distance between the center of the circles is 14.


## Straight-to-Straight Connector

The process is the same as the curve connector, but the width of the rectangle is 25 and the distance between the center of the circles is 20.

# 11

# Airplane

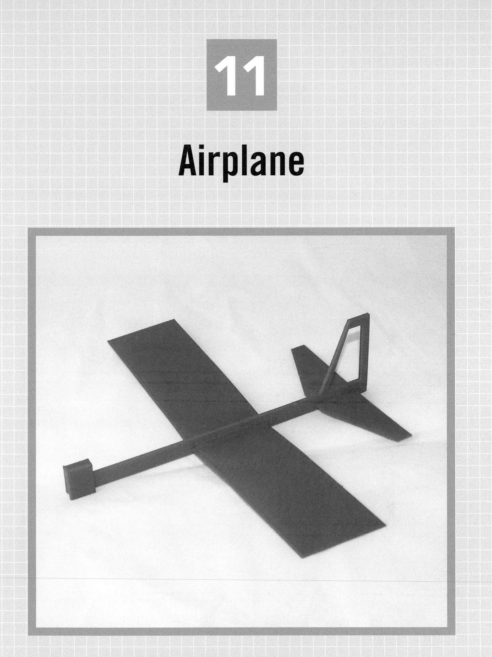

In this chapter, you will design an airplane. There will be a body, wing, and tail, and a weight for the nose.

## Airplane Body

Start by going to **Primitives** and drop down to **Box**.

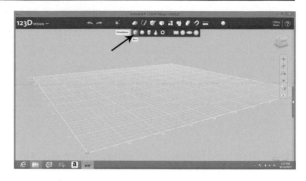

Left-click, release, and move the box to the lower left portion of the screen.

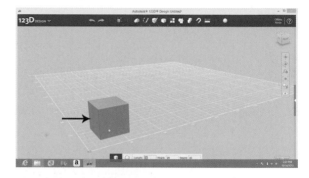

Type 2.5, then press **Tab**. Type 200, then press **Tab**. Type 55.

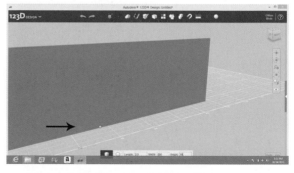

Press **Enter**. Move the cursor to **Front** on the view cube.

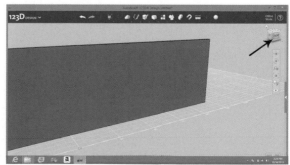

Left-click and release.
Move the cursor to
**Zoom** (third down on
the right).

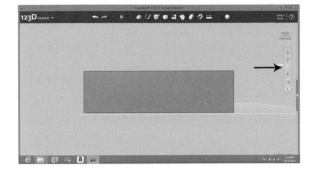

Left-click and release.
Move the cursor into
the work area.

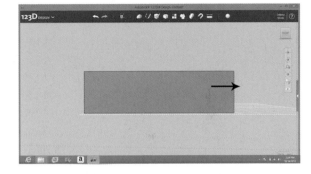

Left-click and hold.
Scroll until the rect-
angle fills most of the
screen. Release the
left-click.

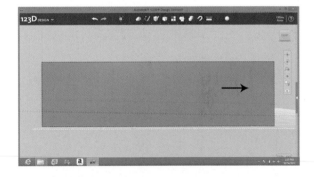

Go to **Sketch** and drop
down to **Polyline**.

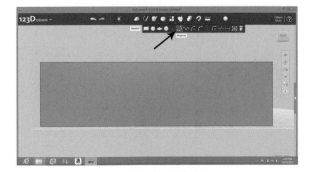

Left-click and release. Move the cursor inside the rectangle.

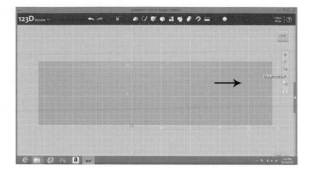

Left-click and release. Move the cursor 3 grid marks left of the upper right corner (15 mm). Left-click and release.

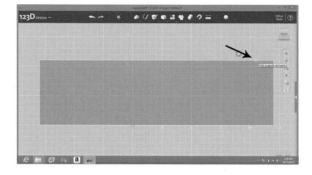

Move the cursor 2 grid marks up and 7 left of the lower right corner (10 mm up, 35 mm left). Left-click and release.

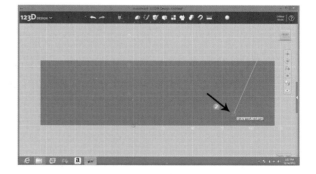

Move the cursor left to the left edge of the rectangle. Left-click and release.

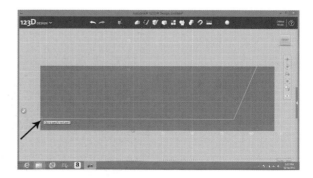

Move the cursor to the upper left corner. Left-click and release.

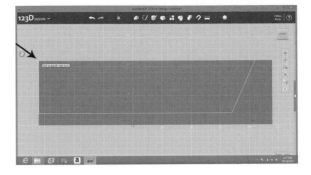

Move the cursor back to the starting point. Left-click and release.

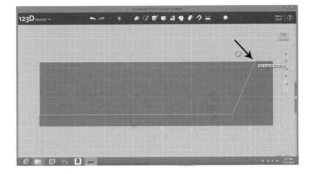

Move the cursor 1 grid mark down and 1 left of the upper right corner. Left-click and release.

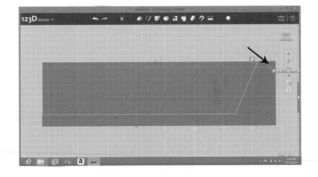

Move the cursor 2 grid marks left. Left-click and release.

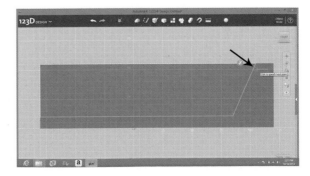

Move the cursor to a point 2 grid marks up and 6 grid marks left of the lower right corner. Left-click and release.

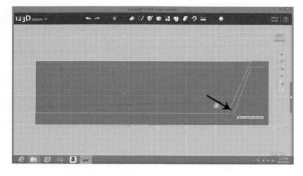

Move the cursor right 5 grid marks. Left-click and release.

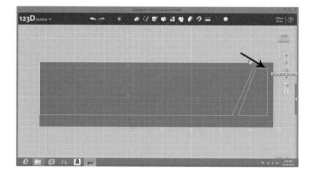

Return the cursor to the spot 1 grid mark down, 1 left of the upper right corner. This will close a polygon. Left-click and release.

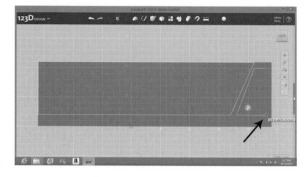

Move the cursor to **Sketch** and drop down to **Rectangle**.

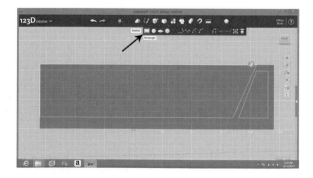

Left-click and release. Move the cursor to the lower part of the plane body.

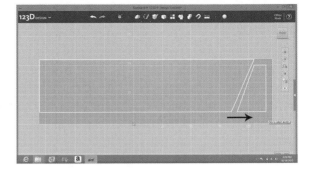

Left-click and release. Move the cursor 1 grid mark up, 1 grid mark left of the lower right corner. Left-click and release. Move the cursor left, then down slightly.

Type 1.5, then press **Tab**. Type 35.

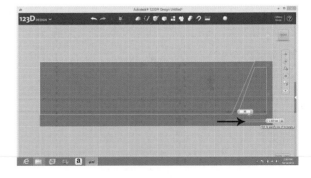

Left-click and release. Move the cursor to a point 1 grid mark up, 14 grid marks left of the lower right corner. Left-click and release. Move the cursor left, then down slightly.

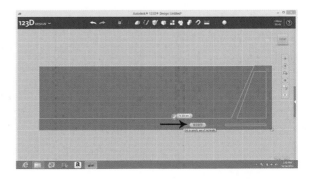

Type 1.5, then press **Tab**. Type 55.

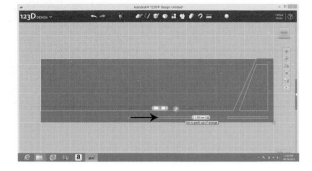

Left-click and release. Go to **Construct**, then drop down to **Extrude**.

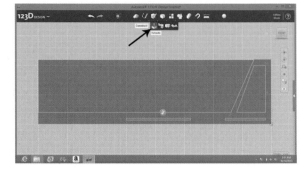

Left-click and release. Move the cursor into the upper left portion of the rectangle. Left-click and release.

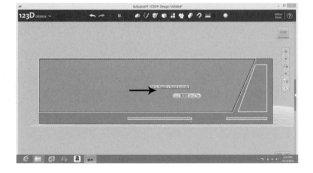

Move the cursor right into the inner part of the airplane tail. Left-click and release.

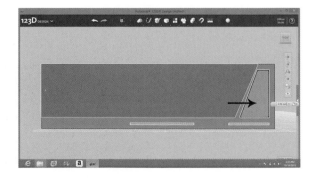

Move the cursor into the thin rectangle at the lower right. Left-click and release.

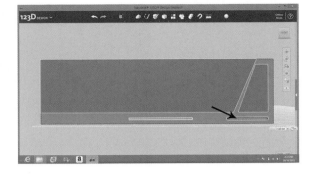

Move the cursor into the thin rectangle in the lower center of the plane. Left-click and release.

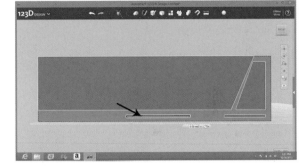

Move the cursor into the white box (to the right end). Left-click and release. Press **Backspace** until the box contains no numbers or letters.

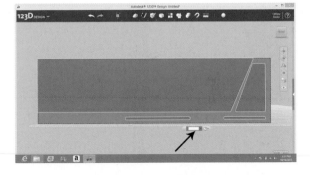

Type -2.5 to remove the highlighted portions.

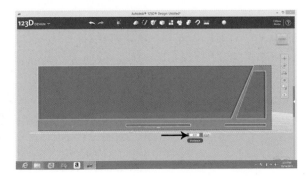

Press **Enter** and move the cursor to **123D** in the upper left corner. On the drop-down menu, select **Export STL** and save your file for printing.

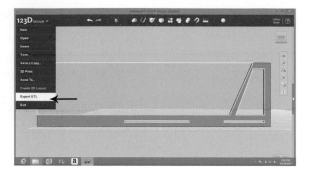

To save the design file, move the cursor to **123D** in the upper left corner and drop down to **Save**.

## Airplane Wing

Go to **Primitives** and drop down to **Box**.

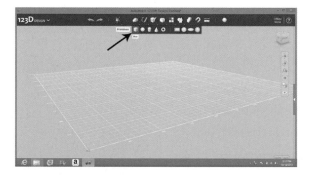

Left-click and release. Move the box into the lower left portion of the screen. Type .6, then press **Tab**. Type 250, then press **Tab**. Type 50.

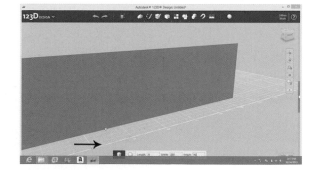

Press **Enter** and move
the cursor to **123D** in
the upper left corner.
On the drop-down
menu, choose **Export
STL** to save your file
for printing.

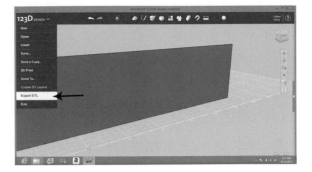

To save the design file, move the cursor to **123D** in the upper left
corner and drop down to **Save**.

## Airplane Tail

Go to **Primitives** and
drop down to **Box**.

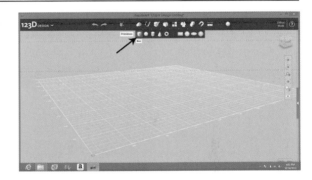

Left-click and release.
Move the box to the
lower left part of the
work area. Type  . 6,
then press **Tab**. Type
120, then press **Tab**.
Type 30.

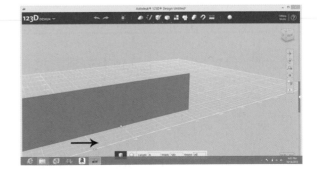

Press **Enter**, then move the cursor to **Front** on the view cube.

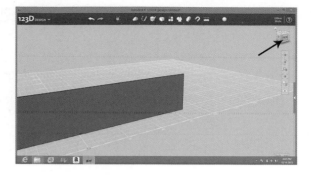

Left-click and release. Move the cursor to **Zoom** on the right side menu.

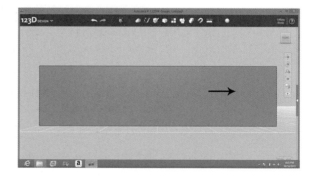

Left-click, release, and move the cursor into the work area. Left-click and hold. Scroll until the rectangle fills most of the work area. Release the left-click.

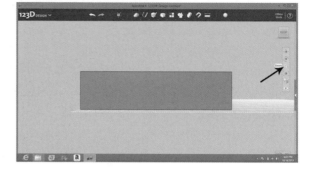

Go to **Sketch** and drop down to **Polyline**. Left-click and release.

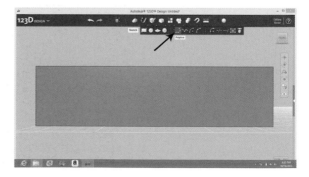

Move the cursor into the rectangle.

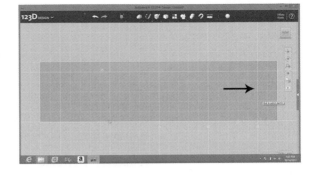

Left-click and release. Move the cursor 2 grid marks above the lower right corner. Left-click and release.

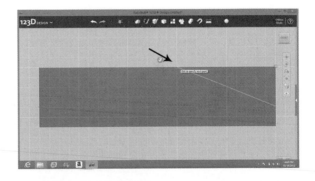

Move the cursor up and left to a point 10 grid marks left of the upper right corner. Left-click and release.

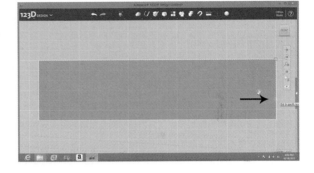

Move the cursor left 4 grid marks. Left-click and release.

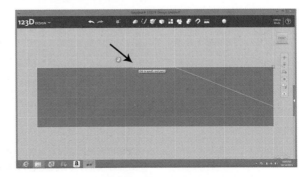

Move the cursor 2 grid marks above the lower left corner. Left-click and release.

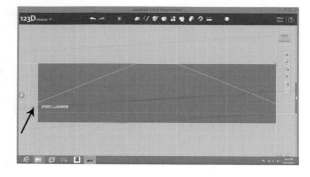

Move the cursor 1 grid mark above the upper left corner. Left-click and release.

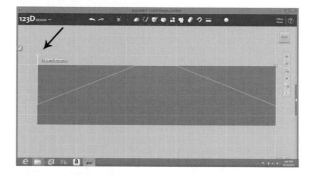

Move the cursor 1 grid mark above the upper right corner. Left-click and release.

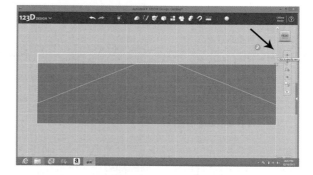

Move the cursor to the starting point (2 grid marks above the lower right corner). Left-click and release.

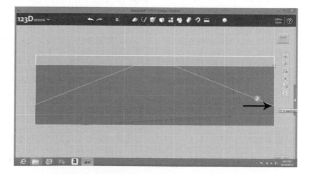

Go to **Construct** and drop down to **Extrude**.

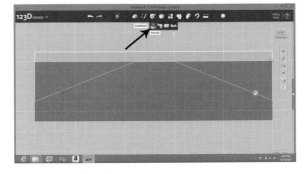

Left-click and release. Move the cursor into the polygon you just created. Left-click and release.

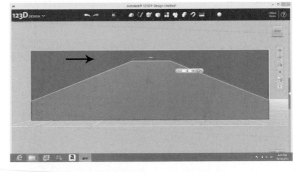

Type − . 6 to remove the highlighted section.

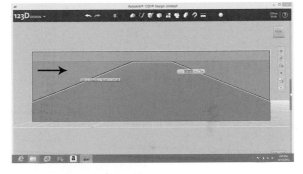

Press **Enter** and move the cursor to **123D** in the upper left corner. On the drop-down menu, select **Export STL** and save your file for printing.

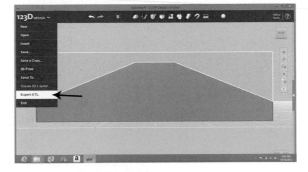

To save the design file, move the cursor to **123D** in the upper left corner and drop down to **Save**.

## Weight for Airplane Nose

Go to **Primitives** and
drop down to **Box**.

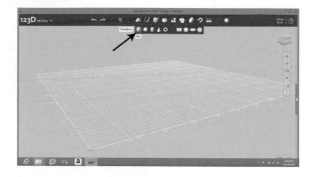

Left-click and release.
Move the box to the
lower left part of the
screen. Type 6, then
press **Tab**. Type 20,
then press **Tab**. Type
15.

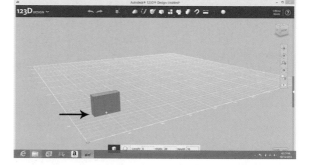

Press **Enter**, then move
the cursor to **Top** on
the view cube.

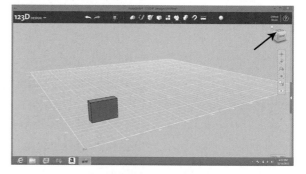

Left-click and release.
Move the cursor to
the **Zoom** icon on the
right menu bar.

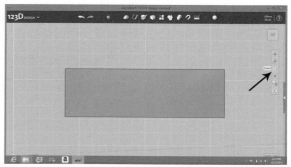

Left-click, release, and move the cursor into the work area. Left-click and hold. Scroll until the rectangle fills most of the work area. Release the left-click.

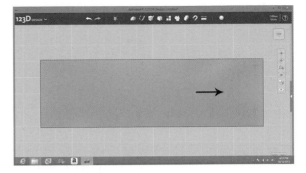

Move the cursor to **Sketch** and drop down to **Rectangle**.

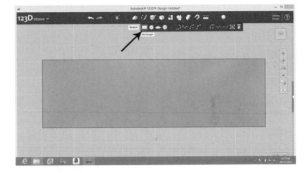

Left-click and release. Move the cursor inside the rectangle.

Left-click and release. Move the cursor to a point 2 mm above the lower left corner. Left-click and release. Move the cursor right and slightly upward.

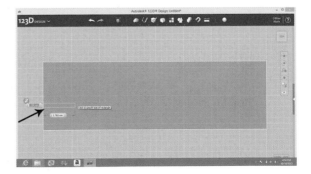

Type 2, then press
**Tab**. Type 15.

Left-click and release.
Move the cursor to
**Construct**, then drop
down to **Extrude**.

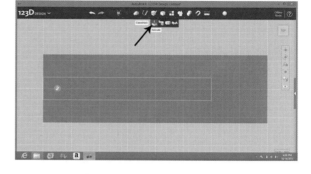

Left-click and release.
Move the cursor inside
the rectangle you just
created. Left-click and
release.

Type −20 to create
a notch.

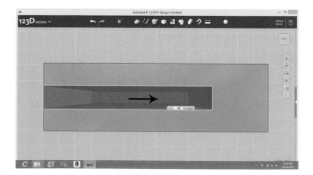

Press **Enter** and move the cursor to **123D** in the upper left corner. On the drop-down menu, select **Export STL** and save your file for printing.

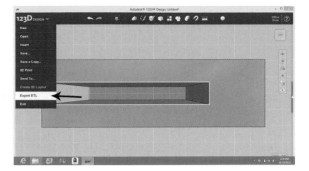

To save the design file, move the cursor to **123D** in the upper left corner and drop down to **Save**.

Insert the tail in the slot at the back of the plane body. Insert the wing in the center slot. Press the weight onto the nose of the plane. Bend the wings up slightly, and let it fly.

If the tail or wing slips around too much, glue it in place.

## MAKING A PLANE BODY USING SKETCHUP

## Airplane Body

Start a new project. Go to **Camera** and drop down to **Standard Views**, then over to **Front**. Left-click and release.

Go to the **Shape/Rectangle** icon. Left-click and release.

Move the pointer to the intersection of red and blue lines. Left-click and hold. Move the cursor right and slightly upward. Type 200,55, then press **Enter**. Release the left-click.

Go to the **Zoom** icon. Left-click and release. Enlarge the rectangle so it fills most of the screen.

Go to the **Tape Measure** icon. Left-click and release.

Move the pointer to the lower left corner. Left-click and hold. Go straight up. Type 10, then press **Enter**. Release the left-click.

Move the pointer to the lower right corner. Left-click and hold. Move the pointer left. Type 5, then press **Enter**. Release the left-click. Move the pointer to the guide point you just created. Left-click and hold. Move the pointer straight up. Type 5, then press **Enter**. Release the left-click.

Move the pointer to the lower right corner. Left-click and hold. Move the pointer left. Type 35, then press **Enter**. Release the left-click. Move the pointer to the guide point you just created. Left-click and hold. Move the pointer straight up. Type 10, then press **Enter**. Release the left-click.

Move the pointer to the lower right corner. Left-click and hold. Move the pointer left. Type 70, then press **Enter**. Release the left-click. Move the pointer to the guide point you just created. Left-click and hold. Move the pointer straight up. Type 5, then press **Enter**. Release the left-click.

Move the pointer to the upper right corner. Left-click and hold. Go horizontally left. Type 15, then press **Enter**. Release the left-click.

Go to **Lines** (third icon from left), then select **Line** from the drop-down menu. Left-click and release. Go to the guide point on the top near the upper right corner. Left-click and hold. Move the pointer to the second guide point from the right. Release the left-click.

With the pointer at this spot, left-click and hold. Move the pointer left until it touches the guide point on the left side of the rectangle. Release the left-click.

Move the pointer into the upper part of the tail. Draw and connect four lines as shown.

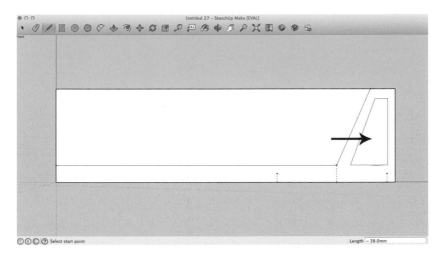

Go to the **Shape/Rectangle** icon. Left-click and release.

Move the pointer to the guide point at the right (5 mm up and 5 mm left of lower right corner). Left-click and hold. Move the pointer left and slightly down. Type 35,1.5, then press **Enter**. Release the left-click.

Move the pointer to the guide point near the center (5 mm up and 70 mm left of lower right corner). Left-click and hold. Move the pointer left and slightly down. Type 55,1.5, then press **Enter**. Release the left-click.

Move to the **Select** icon. Left-click and release. One at a time, select the areas to be removed as indicated in the screen shot on the next page, then choose **Edit**, and drop down to **Delete**, followed by a left-click and release.

You may have to zoom in on the thin rectangles to select and delete them.

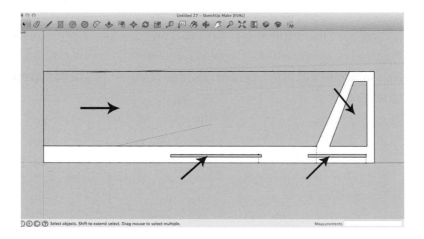

Go to the **Push/Pull** icon. Left-click and release. Move the cursor inside the white plane body. Left-click and hold. Move the cursor upward. Type 2.5, then press **Enter**. Release the left-click.

Move the cursor to the **Select** icon. Left-click and release. Go to **Edit**, then drop down to **Select All**. Left-click and release.

Go to the menu at the top and choose **File**, then drop down to **Export STL**. Left-click. This file can be used to print the plane body.

To save the design file, choose **File**, then **Save**.

## Airplane Wing

Start a new project. Go to **Camera** and drop down to **Standard Views**, then over to **Front**. Left-click and release.

Go to the **Shape/Rectangle** icon. Left-click and release.

Move the pointer to the intersection of red and blue lines. Left-click and hold. Move the cursor right and slightly upward. Type 250,50, then press **Enter**. Release the left-click.

Go to the **Zoom** icon. Left-click and release. Enlarge the rectangle so it fills most of the screen.

Go to the **Push/Pull** icon. Left-click and release. Move the cursor inside the rectangle. Left-click and hold. Move the cursor up slightly. Type `.6`, then press **Enter**. Release the left-click.

Move the cursor to the **Select** icon. Left-click and release. Go to **Edit**, then drop down to **Select All**. Left-click and release.

Go to the menu at the top and choose **File**, then drop down to **Export STL**. Left-click. This file can be used to print the plane wing.

To save the design file, choose **File**, then **Save**.

## Airplane Tail

Start a new project. Go to **Camera** and drop down to **Standard Views**, then over to **Front**. Left-click and release.

Go to the **Shape/Rectangle** icon. Left-click and release.

Move the pointer to the intersection of red and blue lines. Left-click and hold. Move the cursor right and slightly upward. Type `120,30`, then press **Enter**. Release the left-click.

Go to the **Zoom** icon. Left-click and release. Enlarge the rectangle so it fills most of the screen.

Go to the **Tape Measure** icon. Left-click and release.

Move the pointer to the lower left corner. Left-click and hold. Go straight up. Type `10`, then press **Enter**. Release the left-click.

Move the pointer to the lower right corner. Left-click and hold. Go straight up. Type `10`, then press **Enter**. Release the left-click.

Move the pointer to the upper left corner. Left-click and hold. Go right. Type `50`, then press **Enter**. Release the left-click.

Move the pointer to the upper right corner. Left-click and hold. Go left. Type `50`, then press **Enter**. Release the left-click.

Go to **Lines** (third icon from left), then select **Line** from the drop-down menu. Left-click and release.

Move to the guide point on the right side of the rectangle. Left-click and hold. Move the cursor to the guide point on the top (right of center). Release the left-click.

Move to the guide point on the left side of the rectangle. Left-click and hold. Move the cursor to the guide point on the top (left of center). Release the left-click.

Go to the **Select** icon. Left-click and release.

Move the cursor inside the upper left triangle. Left-click and release. Press and hold the **Shift** key. Move the cursor inside the upper right triangle. Left-click and release. Release the **Shift** key.

Go to **Edit** and drop down to **Delete**. Left-click and release.

Go to the **Push/Pull** icon. Left-click and release.

Move the cursor inside the white polygon. Left-click and hold. Move the cursor up slightly. Type . 6, then press **Enter**. Release the left-click.

Move the cursor to the **Select** icon. Left-click and release. Go to **Edit**, then drop down to **Select All**. Left-click and release.

Go to the menu at the top and choose **File**, then drop down to **Export STL**. Left-click. This file can be used to print the plane's tail.

To save the design file, choose **File**, then **Save**.

## Weight for Airplane Nose

Start a new project. Go to **Camera** and drop down to **Standard Views**, then over to **Front**. Left-click and release.

Go to the **Shape/Rectangle** icon. Left-click and release.

Move the pointer to the intersection of red and blue lines. Left-click and hold. Move the cursor right and slightly upward. Type 20, 6, then press **Enter**. Release the left-click.

Go to the **Zoom** icon. Left-click and release. Enlarge the rectangle so it fills most of the screen.

Go to the **Tape Measure** icon. Left-click and release.

Move the pointer to the lower left corner. Left-click and hold. Move the pointer straight up. Type 2, then press **Enter**. Release the left-click.

Go to the **Shape/Rectangle** icon. Left-click and release.

Move the cursor to the guide point. Left-click and hold. Move the cursor right and slightly up. Type 15, 2, then press **Enter**. Release the left-click.

Go to the **Select** icon. Left-click and release.

Move the cursor inside the smaller rectangle. Left-click and release. Go to **Edit**, then drop down to **Delete**. Left-click and release.

Go to the **Push/Pull** icon. Left-click and release.

Move the cursor inside the white area. Left-click and hold. Raise the cursor slightly. Type 15, then press **Enter**. Release the left-click.

Move the cursor to the **Select** icon. Left-click and release. Go to **Edit**, then drop down to **Select All**. Left-click and release.

Go to the menu at the top and choose **File**, then drop down to **Export STL**. Left-click. This file can be used to print the weight for the airplane's nose.

To save the design file, choose **File**, then **Save**.

# 12

# Paddleboat

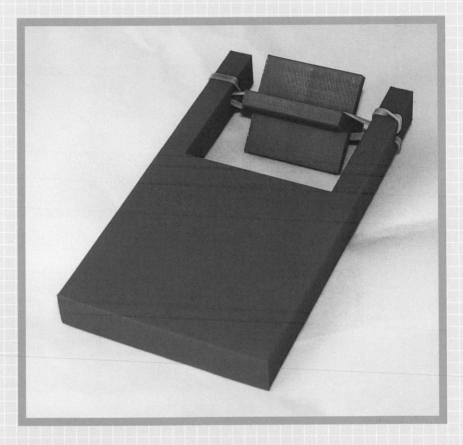

In this chapter you will design a rubber band–powered boat. The body will be a floating structure with space for the paddle wheel and holes to support the rubber band. The paddle wheel will fit into the boat and easily attach to the rubber band.

## Boat

Start by going to **Primitives** and drop down to **Box**.

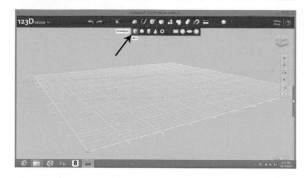

Left-click, release, and move the box to the lower left corner of the screen. Type 15, then press **Tab**. Type 150, then press **Tab**. Type 80.

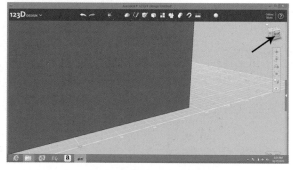

Press **Enter**, then move the cursor to **Front** on the view cube.

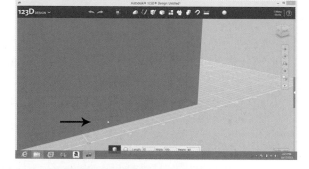

Left-click and release. Go to **Sketch** and drop down to **Rectangle**.

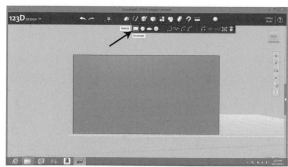

Left-click and release. Move the cursor into the rectangle.

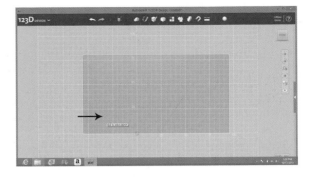

Left-click and release. Move the cursor to a point 2 grid marks above the lower left corner. Left-click and release. Move the cursor right and slightly upward.

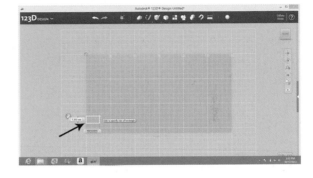

Type 60, then press **Tab**. Type 60.

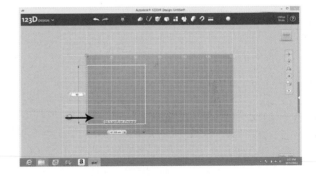

Left-click and release. Move the cursor to **Construct**, then drop down to **Extrude**.

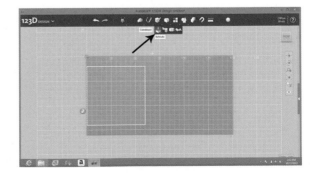

Left-click and release. Move the cursor inside the square. Left-click and release.

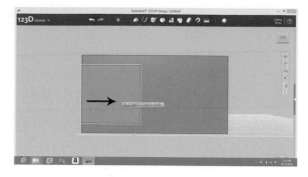

Type −15 to create the space where the paddle wheel will be installed.

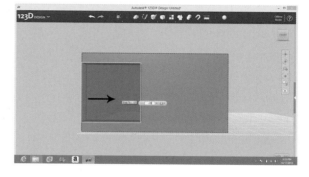

Press **Enter**, then move the cursor to **Top** on the view cube.

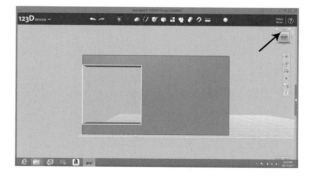

Left-click and release. Go to **Sketch**, then drop down to **Circle**.

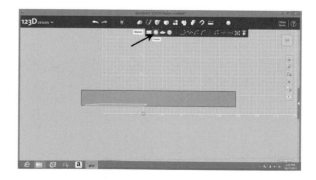

Left-click and release. Move the cursor into the rectangle.

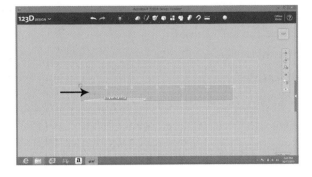

Left-click and release. Move the cursor 4 grid marks right and 1.5 grid marks up from the lower left corner. Left-click and release. Move the cursor slightly.

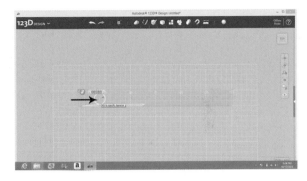

Type 6.

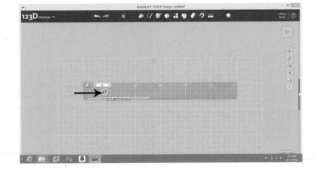

Left-click and release. Go to **Construct**, then drop down to **Extrude**.

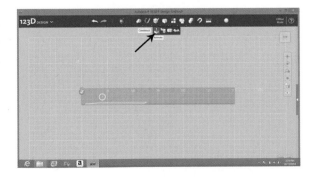

Left-click and release.
Move the cursor inside
the circle. Left-click
and release.

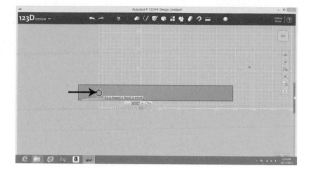

Type −80.

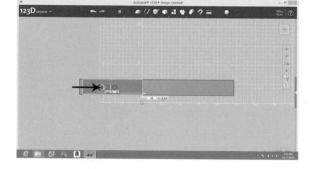

Press **Enter** and move
the cursor to **123D** in
the upper left corner.
On the drop-down
menu, choose **Export
STL** and save your file
for printing.

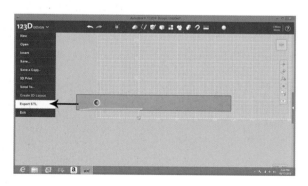

To save the design file, move the cursor to **123D** in the upper left
corner and drop down to **Save**.

# Paddle Wheel

Go to **Primitives** and drop down to **Box**.

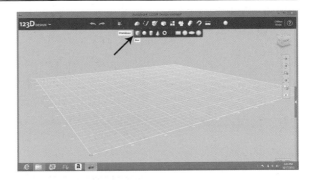

Left-click and release. Move the box to the lower left corner. Type 40, then press **Tab**. Type 40, then press **Tab** again. Type 40.

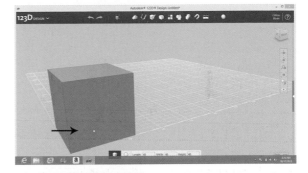

Press **Enter**. Move the cursor to **Front** on the view cube.

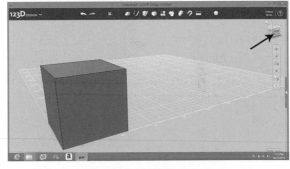

Left-click and release. Move to **Sketch**, then drop down to **Rectangle**.

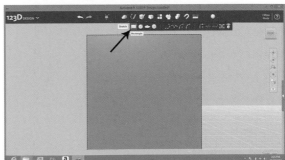

Left-click and release. Move the cursor inside the square.

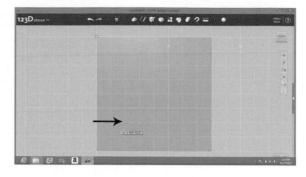

Left-click and release. Move the cursor to the lower left corner. Left-click and release. Move the cursor slightly right and up.

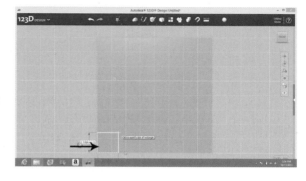

Type 17, then press **Tab**. Type 17.

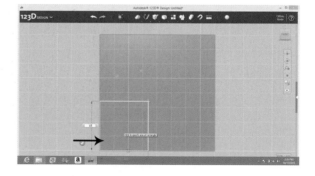

Left-click and release. Move the cursor to the lower right corner. Left-click and release. Move the cursor slightly left and up.

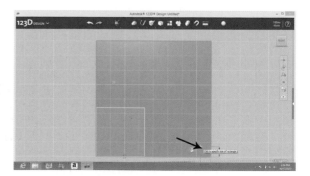

Type 17, then press
**Tab**. Type 17.

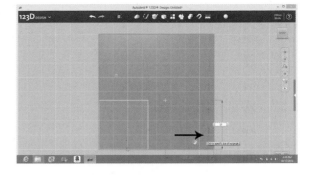

Left-click and release.
Move the cursor to
the upper right corner.
Left-click and release.
Move the cursor
slightly left and down.

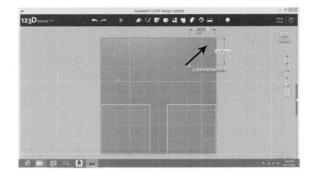

Type 17, then press
**Tab**. Type 17.

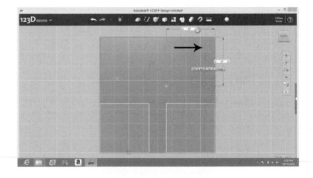

Left-click and release.
Move the cursor to the
upper left corner. Left-
click and release. Move
the cursor slightly right
and down.

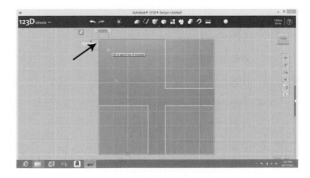

Type 17, then press
**Tab**. Type 17.

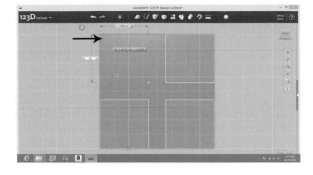

Left-click and release.
Go to **Construct** and
drop down to **Extrude**.

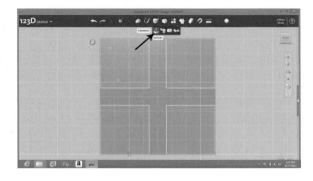

Left-click and release.
Move the cursor into
the upper left square.
Left-click and release.

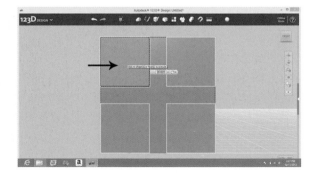

Move the cursor into
the lower left square.
Left-click and release.

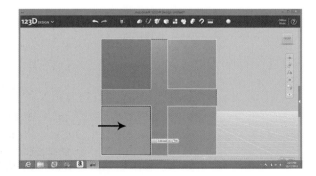

Move the cursor into the lower right square. Left-click and release.

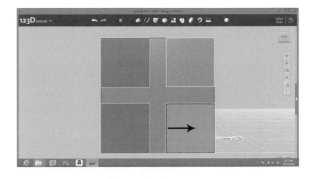

Move the cursor into the upper right square. Left-click and release.

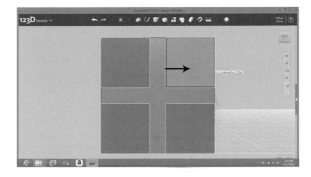

Move the cursor into the small white box (right end). Left-click and release. Press **Backspace** until the numbers and letters are gone.

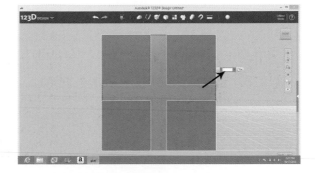

Type −40.

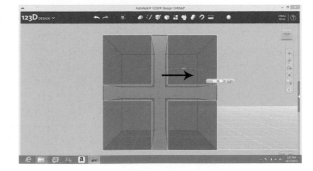

Press **Enter** and move the cursor to **123D** in the upper left corner. On the drop-down menu, choose **Export STL** and save your file for printing.

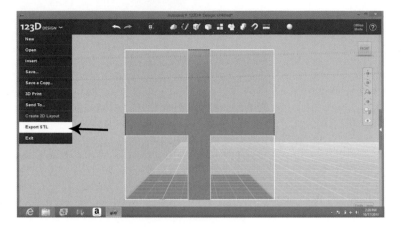

To save the design file, move the cursor to **123D** in the upper left corner and drop down to **Save**.

Hook a rubber band around one of the boat legs, pass it through both holes, and hook it on the other boat leg. Pull the rubber band around the paddle. Wind it up and enjoy power boating!

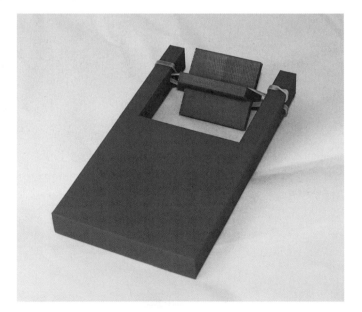

# MAKING A PADDLEBOAT USING SKETCHUP

## Boat

Start a new project. Go to **Camera**, then drop down to **Standard Views**, then over to **Top**. Left-click and release.

Go to the **Shape/Rectangle** icon. Left-click and release.

Move the pointer to the intersection of red and green lines. Left-click and hold. Move the cursor right and slightly upward. Type 150,80, then press **Enter**. Release the left-click.

Go to the **Zoom Extents** icon. Left-click and release.

Go to the **Tape Measure** icon. Left-click and release.

Go to the lower right corner. Left-click and hold. Move the cursor straight upward. Type 10, then press **Enter**. Release the left-click.

Go to the **Shape/Rectangle** icon. Left-click and release.

Move the pointer to the guide point you just created. Left-click and hold. Move the pointer left and up. Type 60,60, then press **Enter**. Release the left-click.

Go to the **Select** icon. Left-click and release. Move the cursor inside the square you just created. Left-click and release. Go to **Edit**, then drop down to **Delete**. Left-click and release.

Go to the **Push/Pull** icon. Left-click and release. Move the cursor into the lower part of the blue area (part of the original rectangle you drew). Left-click and hold. Move the cursor upward. Type 15, then press **Enter**. Release the left-click.

Go to **Camera**, then drop down to **Standard Views**. Go to **Front**, then left-click and release.

Go to the **Zoom Extents** icon. Left-click and release.

Go to the **Tape Measure** icon. Left-click and release. Go to the lower right corner. Left-click and hold. Move the cursor horizontally left. Type 20, then press **Enter**. Release the left-click.

Move the pointer to the guide point you just created. Left-click and hold. Move the cursor upward. Type 7.5, then press **Enter**. Release the left-click.

Go to the **Shape/Circle** icon. Left-click and release. Move the pointer to the guide point. Left-click and hold. Move the cursor slightly. Type 3, then press **Enter**. Release the left-click.

Go to the **Push/Pull** icon. Left-click and release. Move the cursor inside the circle. Left-click and hold. Move the cursor to the left. Type 10, then press **Enter**. Release the left-click.

Go to **Camera**, the drop down to **Standard Views**. Choose **Back**, then left-click and release.

Go to the **Tape Measure** icon. Left-click and release. Go to the lower left corner. Left-click and hold. Move the cursor horizontally right. Type 20, then press **Enter**. Release the left-click.

Move the pointer to the guide point you just created. Left-click and hold. Move the cursor upward. Type 7.5, then press **Enter**. Release the left-click.

Go to the **Shape/Circle** icon. Left-click and release. Move the pointer to the guide point. Left-click and hold. Move the cursor slightly. Type 3, then press **Enter**. Release the left-click.

Go to the **Push/Pull** icon. Left-click and release. Move the cursor inside the circle. Left-click and hold. Move the cursor to the right. Type 10, then press **Enter**. Release the left-click.

Move the cursor to the **Select** icon. Left-click and release. Go to **Edit**, then drop down to **Select All**. Left-click and release.

Go to the menu at the top and choose **File**, then drop down to **Export STL**. Left-click. This file can be used to print the boat body.

To save the design file, choose **File**, then **Save**.


## Paddle Wheel

Start a new project. Go to **Camera** and drop down to **Standard Views**, then over to **Top**. Left-click and release.

Go to the **Shape/Rectangle** icon. Left-click and release.

Move the pointer to the intersection of red and green lines. Left-click and hold. Move the cursor right and slightly upward. Type 40,40, then press **Enter**. Release the left-click.

Go to the **Zoom** icon. Left-click and release. Scroll until the square fills about half the screen.

Go to the **Shape/Rectangle** icon. Left-click and release.

Move the pointer to the upper left corner. Left-click and hold. Move the cursor right and down. Type 17,17, then press **Enter**. Release the left-click.

Move the pointer to the upper right corner. Left-click and hold. Move the cursor left and down. Type 17,17, then press **Enter**. Release the left-click.

Move the pointer to the lower left corner. Left-click and hold. Move the cursor right and up. Type 17,17, then press **Enter**. Release the left-click.

Move the pointer to the lower right corner. Left-click and hold. Move the cursor left and up. Type 17,17, then press **Enter**. Release the left-click.

Go to the **Select** icon. Left-click and release.

Move the cursor inside the upper left square. Left-click and release.

Press the **Shift** key and hold. Move the cursor into the upper right square. Left-click and release the left-click. Move the cursor into the lower right square. Left-click and release the left-click. Move the cursor into the lower left square. Left-click and release the left-click. Release the **Shift** key.

Go to **Edit**, then drop down to **Delete**. Left-click and release.

Go to the **Push/Pull** icon. Left-click and release.

Move the cursor into the blue cross. Left-click and hold. Move the cursor slightly. Type 40, then press **Enter**. Release the left-click.

Move the cursor to the **Select** icon. Left-click and release. Go to **Edit**, then drop down to **Select All**. Left-click and release.

Go to the menu at the top and choose **File**, then drop down to **Export STL**. Left-click. This file can be used to print the paddle wheel.

To save the design file, choose **File**, then **Save**.

# 13

# Drum

To make a drum, you first create a cylinder, carve out a hole, then stretch a balloon over the top.

First, go to **Primitives**, then drop down to **Cylinder**.

Left-click and release. Move the cylinder to the lower left portion of the work area.

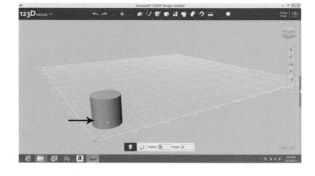

Type 55 for the radius, then press **Tab**. Type 90 for the height.

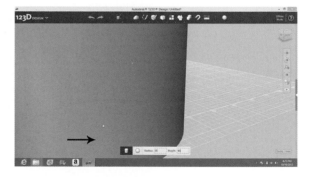

Press **Enter**, then move the cursor to **Top** on the view cube.

Left-click and release.
Go to **Sketch**, then
drop down to **Circle**.

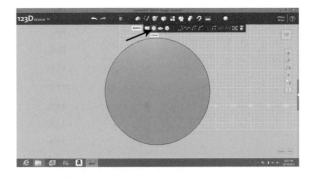

Left-click and release.
Move the cursor into
the circle.

Left-click and release.
Go to the center of the
circle (11 grid marks
up, down, left and
right of this point).
Left-click and release.
Move the cursor
slightly.

Type 100 for the
diameter.

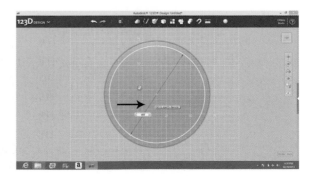

Left-click and release. Go to **Construct**, then drop down to **Extrude**.

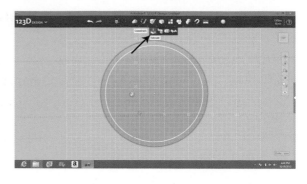

Left-click and release. Move the cursor inside the center circle. Left-click and release.

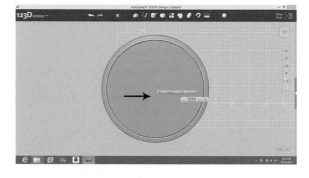

Type -85 to carve out the inside of the drum.

Press **Enter** and move the cursor to **123D** in the upper left corner. On the drop-down menu, select **Export STL** and save your file for printing.

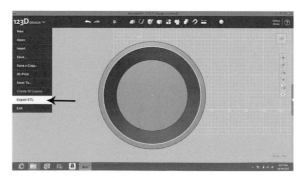

To save the design file, move the cursor to **123D** in the upper left corner and drop down to **Save**.

Using scissors, cut off the narrow nozzle end of a 12-inch balloon. Stretch the remaining part of the balloon over the open end of the drum. Use the eraser end of a pencil as a drumstick and enjoy your new instrument.

## MAKING A DRUM USING SKETCHUP

Start a new project. Go to **Camera** and drop down to **Standard Views**, then over to **Top**. Left-click and release.

Go to the **Shape/Circle** icon. Left-click and release.

Move the pointer to the intersection of red and green lines. Left-click and hold. Move the cursor right. Type 55, then press **Enter**. Release the left-click.

Go to the **Zoom Extents** icon. Left-click and release.

Go to the **Push/Pull** icon. Left-click and release. Move the cursor inside the left half of the circle. Left-click and hold. Move the cursor slightly to the right. Type 90, then press **Enter**. Release the left-click.

Go to the **Shape/Circle** icon. Left-click and release.

Move the pointer to the intersection of red and green lines. Left-click and hold. Move the cursor right. Type 50, then press **Enter**. Release the left-click.

Go to the **Push/Pull** icon. Left-click and release. Move the cursor inside the left half of the inner circle. Left-click and hold. Move the cursor slightly to the right. Type 85, then press **Enter**. Release the left-click.

Move the cursor to the **Select** icon. Left-click and release. Go to **Edit**, then drop down to **Select All**. Left-click and release.

Go to the menu at the top and choose **File**, then drop down to **Export STL**. Left-click. This file can be used to print the drum.

To save the design file, choose **File**, then **Save**.

# 14

# Rubber Band Car

In this chapter, you will design a rubber band–powered car. The car is made from a base, two sides, and four wheels.

## Base

Go to **Primitives** and drop down to **Box**.

Left-click and release. Move the box to the lower left corner of the work area. Type 80, then press **Tab**. Type 150, then press **Tab**. Type 5.

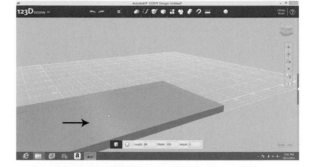

Press **Enter**, then move the cursor to **Front** on the view cube.

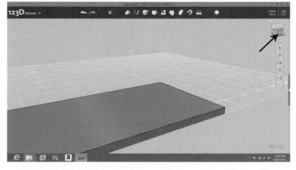

Left-click and release. Go to **Sketch**, then drop down to **Rectangle**.

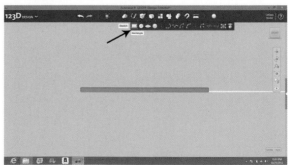

Left-click and release.
Move the cursor inside
the rectangle.

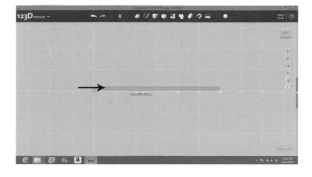

Left-click and release.
Move the cursor
25 mm right and
5 mm up from the
lower left corner. Left-
click and release. Move
the cursor right and
slightly down.

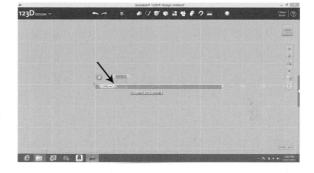

Type 100, then press
**Tab**. Type 5.

Left-click and release.
Move the cursor to
**Construct** and drop
down to **Extrude**.

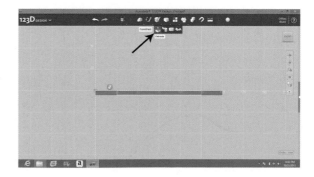

Left-click and release. Move the cursor inside the inner rectangle. Left-click and release.

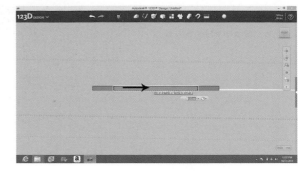

Type −85 to add material in a downward direction.

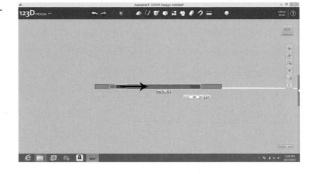

Move the cursor to the small black arrow at the far right of the box where the typed number appeared. Left-click and release. Move the cursor down to **Merge**. Left-click and release.

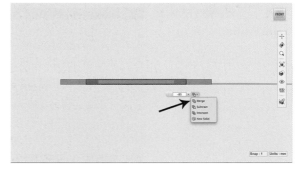

Press **Enter**. This creates a protruding section on the back side of the base.

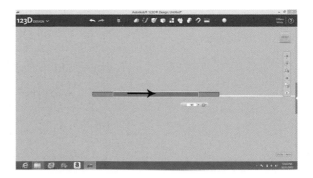

Go to **Construct**, then drop down to **Extrude**.

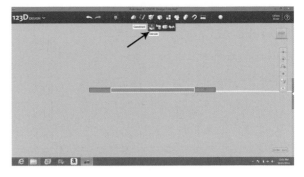

Left-click and release. Move the cursor inside the inner rectangle. Left-click and release.

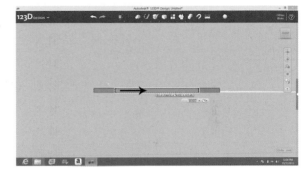

Type 5. This will create a protruding section on the front side of the base.

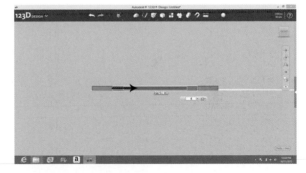

Press **Enter**. Move the cursor to **Left** on the view cube.

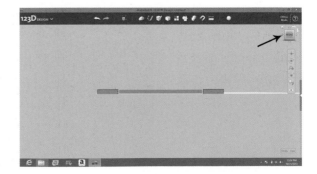

Left-click and release. Move the cursor to **Zoom** on the right side menu bar.

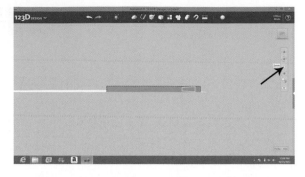

Left-click and release. Move the cursor into the work area. Left-click and hold while scrolling until the rectangle fills most of the screen. Release the left-click. Go to **Sketch** and drop down to **Rectangle**.

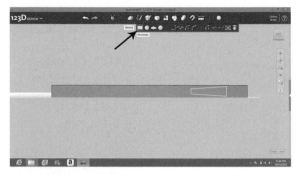

Left-click and release. Move the cursor into the rectangle. Left-click and release. Move the cursor 7 grid marks right from the lower left corner.

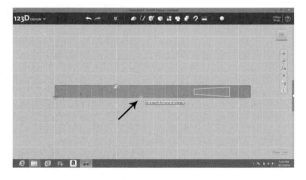

Left-click and release. Move the cursor right and upward. Type 10, then press **Tab**. Type 5.

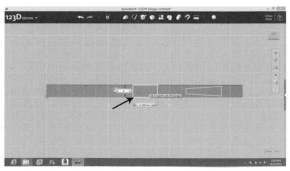

Left-click and release. Go to **Construct** and drop down to **Extrude**.

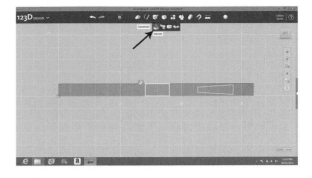

Left-click and release. Move the cursor inside the small rectangle. Left-click and release.

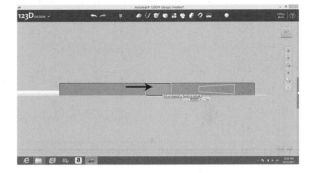

Type 5.

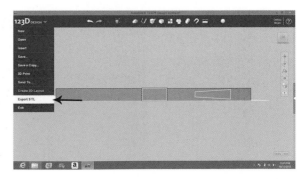

Press **Enter** and move the cursor to **123D** in the upper left corner. On the drop-down menu, select **Export STL** and save your file for printing.

To save the design file, move the cursor to **123D** in the upper left corner and drop down to **Save**.

## Side

Start a new project. Go to **Primitives** and drop down to **Box**.

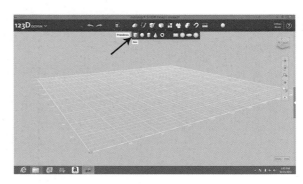

Left-click and release. Move the box into the lower left part of the work area. Type 5, then press **Tab**. Type 150, then press **Tab**. Type 40.

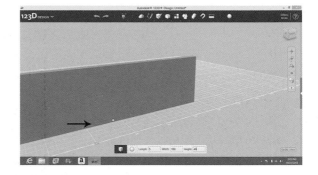

Press **Enter** and move the cursor to **Front** on the view cube.

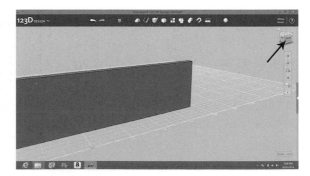

Left-click and release. Go to **Sketch** and drop down to **Rectangle**.

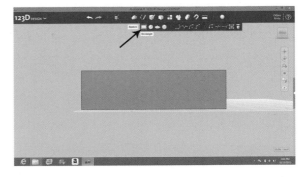

Left-click and release. Move the cursor inside the rectangle.

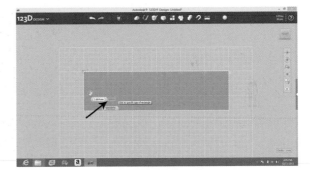

Left-click and release. Move the cursor 5 grid marks right and 2 grid marks up from the lower left corner. Left-click and release. Move the cursor right and slightly upward.

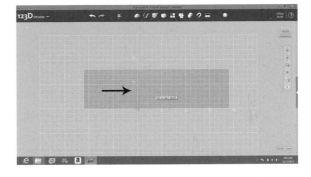

Type 5 . 4, then press **Tab**. Type 101. These dimensions are slightly larger than the 100 mm by 5 mm base piece this side piece will have to slide onto.

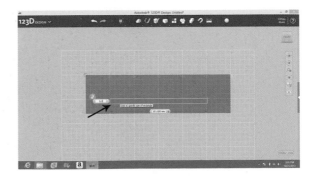

Left-click and release.
Go to **Sketch** and drop
down to **Circle**.

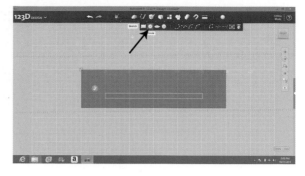

Left-click and release.
Move the cursor inside
the rectangle.

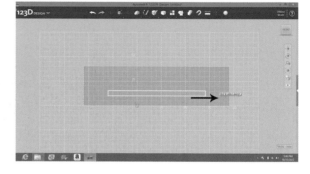

Left-click and release.
Move the cursor to a
point 5 grid marks up
and 2 grid marks left
from the lower right
corner. Left-click and
release. Move the cur-
sor slightly.

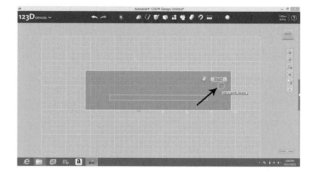

Type 8 . 5.

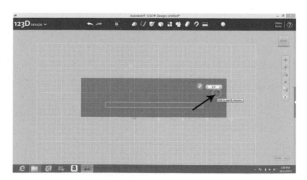

Left-click and release. Move the cursor to a point 5 grid marks up and 2 grid marks right from the lower left corner. Left-click and release. Move the cursor slightly.

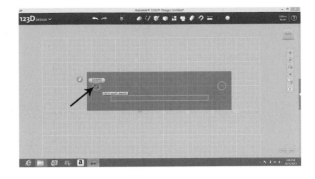

Type 8 . 5 .

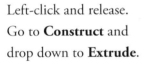

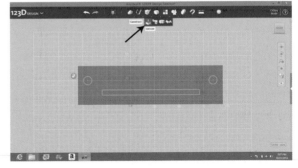

Left-click and release. Go to **Construct** and drop down to **Extrude**.

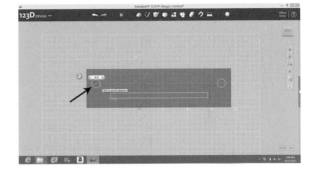

Left-click and release. Move the cursor into the circle on the left. Left-click and release.

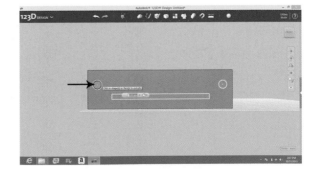

Move the cursor into the circle on the right. Left-click and release.

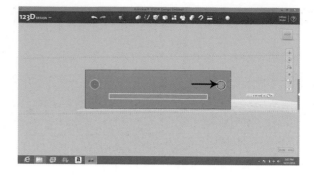

Move the cursor into the thin rectangle below the two circles. Left-click and release.

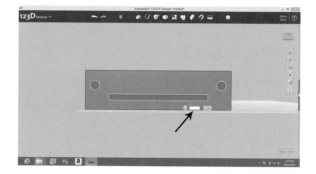

Move the cursor into the white box (right end). Left-click and release. Press **Backspace** until all the letters and numbers are gone.

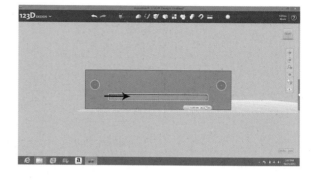

Type −5 to create the holes where the axles and the base fit into the side piece.

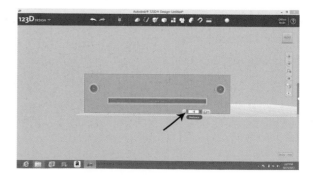

Press **Enter** and move
the cursor to **123D** in
the upper left corner.
On the drop-down
menu, select **Export
STL** and save your file
for printing.

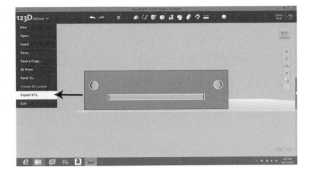

To save the design file, move the cursor to **123D** in the upper left corner and drop down to **Save**.

## Wheel

Go to **Primitives**
and drop down to
**Cylinder**.

Left-click and release.
Move the cylinder to
the lower left part of
the work area. Type
55, then press **Tab**.
Type 5.

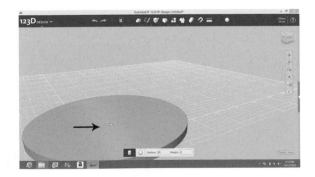

Press **Enter**, then move the cursor to **Top** on the view cube.

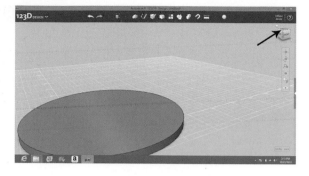

Left-click and release. Move to **Sketch** and drop down to **Circle**.

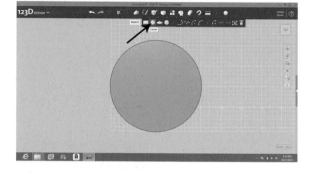

Left-click and release. Move the cursor into the circle. Left-click and release. Go to the center of the circle (11 grid marks from the top, bottom, left, and right edges). Left-click and release. Move the cursor slightly.

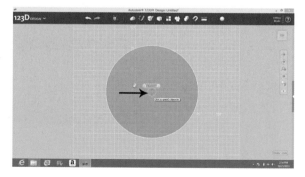

Type 7 . 7. This is the diameter of the pencil that will be used for an axle.

Left-click and release. Go to **Construct** and drop down to **Extrude**.

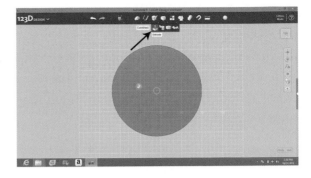

Left-click and release. Move the cursor inside the small center circle. Left-click and release.

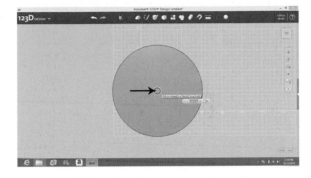

Type −5.

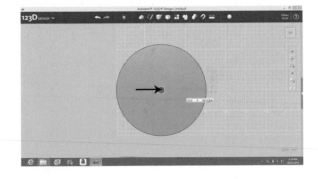

Press **Enter** and move the cursor to **123D** in the upper left corner. On the drop-down menu, choose **Export STL** to save your file for printing.

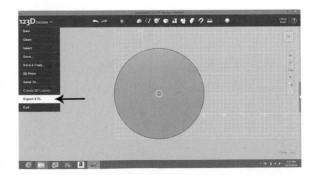

To save the design file, move the cursor to **123D** in the upper left corner and drop down to **Save**.

Push the two sides onto the base. Insert a pencil through one wheel, then the pencil/wheel assembly through the rear part of the base assembly. Attach the other wheel. Repeat for the front. Attach a rubber band, then wind it up and release.

The rubber band car is adapted from the book *Amazing Rubber Band Cars* by Mike Rigsby. Patterns for other cardboard toys are available in that book.

# MAKING A RUBBER BAND CAR USING SKETCHUP

## Base

Start a new project. Go to **Camera**, then drop down to **Standard Views**. Go to **Top**, then left-click and release.

Go to the **Shape/Rectangle** icon. Left-click and release.

Move the pointer to the intersection of red and green lines. Left-click and hold. Move the cursor right and slightly upward. Type 150,90, then press **Enter**. Release the left-click.

Go to the **Zoom** icon. Left-click and release. Move the cursor to the rectangle, then scroll until the rectangle fills two-thirds of the screen.

Go to the **Tape Measure** icon. Left-click and release.

Move the pointer to the lower left corner. Left-click and hold. Move the pointer upward. Type 40, then press **Enter**. Release the left-click.

Go to the **Shape/Rectangle** icon. Left-click and release.

Move the pointer until it rests on the guide point. Left-click and hold. Move the cursor up and slightly left. Type 10,5, then press **Enter**. Release the left-click.

Go to the **Push/Pull** icon. Left-click and release.

Move the cursor to the small rectangle you just created. (It should fill with small black dots.) Left-click and hold. Move the cursor to the right. Type 5, then press **Enter**. Release the left-click.

Move the cursor inside the large rectangle. Left-click and hold. Move the cursor to the right. Type 5, then press **Enter**. Release the left-click.

Go to the **Shape/Rectangle** icon. Left-click and release.

Go to the upper left corner. Left-click and hold. Move the cursor right and slightly down. Type 25,5, then press **Enter**. Release the left-click.

Go to the lower left corner. Left-click and hold. Move the cursor right and slightly up. Type 25,5, then press **Enter**. Release the left-click.

Go to the upper right corner. Left-click and hold. Move the cursor left and slightly down. Type 25, 5, then press **Enter**. Release the left-click.

Go to the lower right corner. Left-click and hold. Move the cursor left and slightly up. Type 25, 5, then press **Enter**. Release the left-click.

Go to the **Push/Pull** icon. Left-click and release.

Move the cursor into the upper left rectangle. Left-click and hold. Move the cursor slightly right. Type 5, then press **Enter**. Release the left-click.

Move the cursor into the lower left rectangle. Left-click and hold. Move the cursor slightly right. Type 5, then press **Enter**. Release the left-click.

Move the cursor into the upper right rectangle. Left-click and hold. Move the cursor slightly left. Type 5, then press **Enter**. Release the left-click.

Move the cursor into the lower right rectangle. Left-click and hold. Move the cursor slightly left. Type 5, then press **Enter**. Release the left-click.

Move the cursor to the **Select** icon. Left-click and release. Go to **Edit**, then drop down to **Select All**. Left-click and release.

Go to the menu at the top and choose **File**, then drop down to **Export STL**. Left-click. This file can be used to print the rubber band car base.

To save the design file, choose **File**, then **Save**.

## Side

Start a new project. Go to **Camera**, then drop down to **Standard Views**. Go to **Top**, then left-click and release.

Go to the **Shape/Rectangle** icon. Left-click and release.

Move the pointer to the intersection of red and green lines. Left-click and hold. Move the cursor right and slightly upward. Type 150, 40, then press **Enter**. Release the left-click.

Go to the **Zoom Extents** icon. Left-click and release.

Go to the **Tape Measure** icon. Left-click and release.

Move the pointer to the lower left corner. Left-click and hold. Move the cursor to the right. Type 25, then press **Enter**. Release the left-click.

Move the pointer to the guide point you just created. Left-click and hold. Move the cursor upward. Type 10, then press **Enter**. Release the left-click.

Move the pointer to the lower left corner. Left-click and hold. Move the cursor straight up. Type 25, then press **Enter**. Release the left-click.

Move the pointer to the guide point you just created. Left-click and hold. Move the cursor to the right. Type 10, then press **Enter**. Release the left-click.

Move the pointer to the lower right corner. Left-click and hold. Move the cursor straight up. Type 25, then press **Enter**. Release the left-click.

Move the pointer to the guide point you just created. Left-click and hold. Move the cursor left. Type 10, then press **Enter**. Release the left-click.

Go to the **Shape/Circle** icon. Left-click and release.

Move the pointer to the guide point that is inside the rectangle (upper left). Left-click and hold. Move the cursor. Type 4.25, then press **Enter**. Release the left-click.

Move the pointer to the guide point inside the rectangle (upper right). Left-click and hold. Move the cursor. Type 4.25, then press **Enter**. Release the left-click.

Go to the **Shape/Rectangle** icon. Left-click and release.

Move the pointer to the guide point inside the rectangle (lower left). Left-click and hold. Move the cursor right and slightly up. Type 101,5.4, then press **Enter**. Release the left-click.

Go to the **Select** icon. Left-click and release.

Move the cursor inside the left circle. Left-click and release. Press and hold the **Shift** key. Move the cursor inside the thin rectangle. Left-click and release the left-click. Move the cursor into the right circle. Left-click and release the left-click. Make sure you only select the area inside the lines. Release the **Shift** key.

Go to **Edit**, then drop down to **Delete**. Left-click and release.

Go to the **Push/Pull** icon. Left-click and release.

Move the cursor inside the blue part of the rectangle. Left-click and hold. Move the cursor to the right. Type 5, then press **Enter**. Release the left-click.

Move the cursor to the **Select** icon. Left-click and release. Go to **Edit**, then drop down to **Select All**. Left-click and release.

Go to the menu at the top and choose **File**, then drop down to **Export STL**. Left-click. This file can be used to print the rubber band car sides.

To save the design file, choose **File**, then **Save**.

## Wheel

Start with a new file. Go to **Camera**, then drop down to **Standard Views**. Go to **Top**, then left-click and release.

Go to the **Shape/Circle** icon. Left-click and release.

Move the pointer to the intersection of red and green lines. Left-click and hold. Move the cursor. Type 55, then press **Enter**. Release the left-click.

Go to the **Zoom Extents** icon. Left-click and release.

Move the pointer to the center of the circle. Left-click and hold. Move the cursor. Type 3.85, then press **Enter**. Release the left-click.

Go to the **Zoom** icon. Left-click and release. Enlarge the center circle so it fills at least 20 percent of the screen. (Your next step, the **Select** icon, will not select an item if it is too small.)

Go to the **Select** icon. Left-click and release. Move the cursor inside the center circle. Left-click and release.

Go to **Edit**, then drop down to **Delete**. Left-click and release.

Go to the **Push/Pull** icon. Left-click and release. Move the cursor into the blue area. Left-click and hold. Move the cursor to the right. Type 5, then press **Enter**. Release the left-click.

Move the cursor to the **Select** icon. Left-click and release. Go to **Edit**, then drop down to **Select All**. Left-click and release.

Go to the menu at the top and choose **File**, then drop down to **Export STL**. Left-click. This file can be used to print the rubber band car wheels.

To save the design file, choose **File**, then **Save**.

# 15

# Little Clicker

Little Clicker is a push toy that makes a loud clicking sound when a paddle wheel (in contact with a business card) is turned. Little Clicker has a body, two wheels, a clicker paddle, and a decorative top with eyeballs.

## Decorative Top

Start a new project. Go to **Primitives** and drop down to **Box**.

Left-click and release. Move the box to the lower left portion of the screen. Type 85, then press **Tab**. Type 140, then press **Tab**. Type 10.

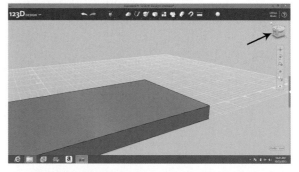

Press **Enter**, then move the cursor to **Left** on the view cube.

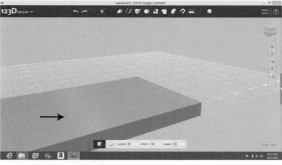

Left-click and release. Go to **Sketch** and drop down to **Three Point Arc**.

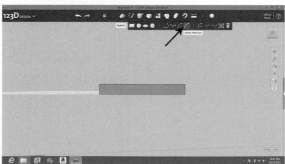

Left-click and release. Move the cursor into the rectangle. Left-click and release.

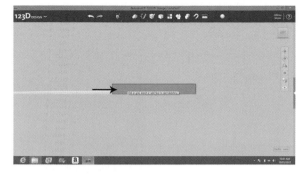

Move the cursor to the upper left corner of the rectangle. Make sure the cursor is on or slightly inside the rectangle.

Left-click and release. Move the cursor horizontally to the top right corner of the rectangle.

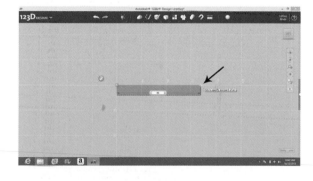

Left-click and release. Move the cursor up until the peak of the arc is about 35 mm above the rectangle.

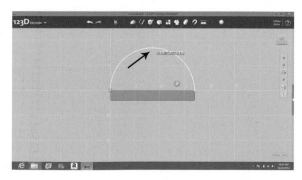

Left-click and release.
Move to **Sketch**,
then drop down to
**Polyline**.

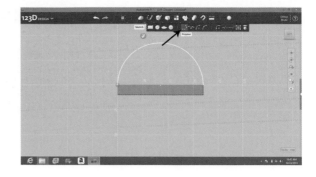

Left-click and release.
Move the cursor to the
left end of the arc. The
arc will become thick
and bold.

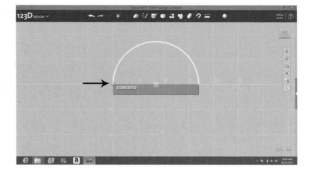

Left-click and release.
Move the cursor to the
right end of the arc.

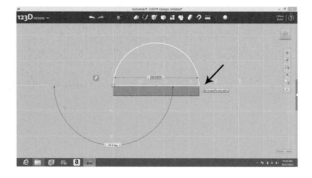

Left-click and release.
Go to **Construct** and
drop down to **Extrude**.
Left-click and release.

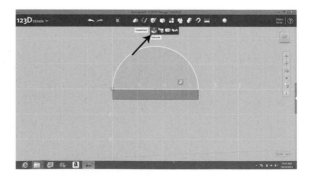

Move the cursor into the semicircle. Left-click and release.

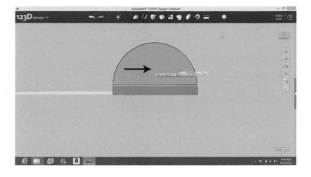

Type -140.

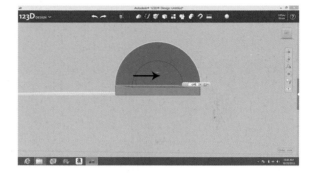

Move the cursor to the small black arrow at the far right of the box where you typed "-140." Left-click and release. Drop down to **Merge**.

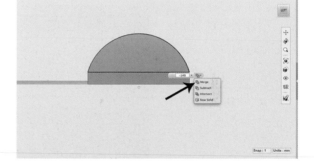

Press **Enter**, then move to **Pan** on the right side menu.

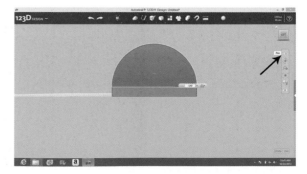

Left-click and release. Move the cursor to the center of the screen. Left-click and hold. Move the assembly you have created to the lower center of the screen.

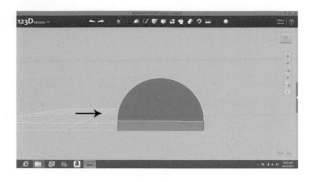

Go to **Sketch** and drop down to **Spline**. Left-click and release.

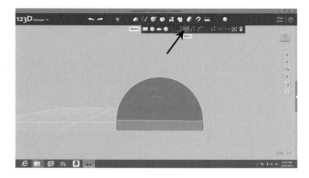

Move the cursor inside the semicircle.

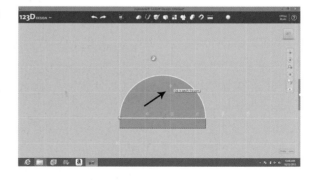

You will be creating a stalk to hold the eyeballs. This is a free-form shape—meaning you can create what-ever you want. Left-click and release. Move the cursor up.

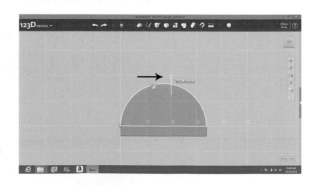

Left-click and release. Move the cursor a bit, then left-click and release. Move the cursor more, then left-click and release. Continue until your shape is done. (I made nine additional points.) The last point must be inside the semicircle.

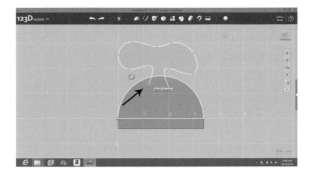

Go to **Construct** and drop down to **Extrude**.

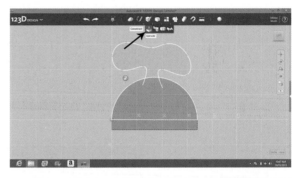

Left-click and release. Move the cursor into the eyestalk you just created. Left-click and release.

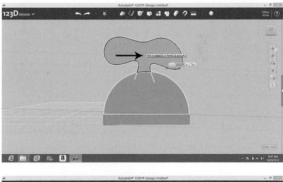

Type -30 to add material in a downward direction.

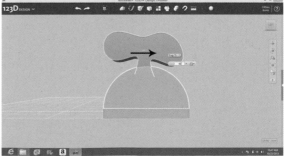

Press **Enter**. Go to
**Construct**, then drop
down to **Extrude**.

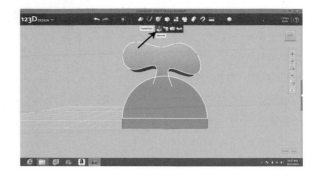

Left-click and release.
Move the cursor into
the eyestalk again.
Left-click and release.

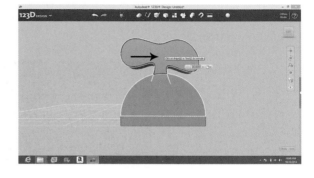

Type -10. This
removes the first
10 mm of the eyestalk,
so it is set back from
the front face.

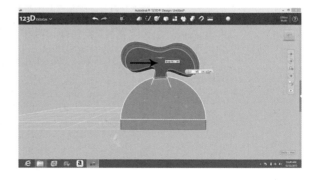

Press **Enter**. Left-click
and release.

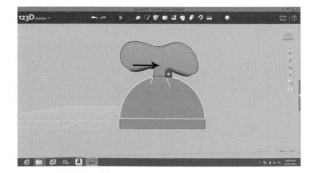

A small gear appears in the lower right corner of the eyestalk. Move the cursor to the gear, and a series of pictures will appear to the right of the gear. Move to the **Hide** icon (far right).

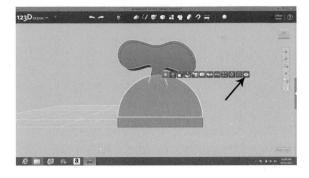

Left-click and release. The "hazy" front of the eyestalk will vanish.

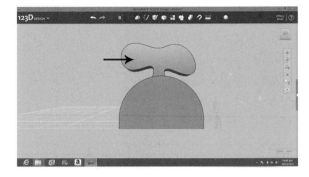

Go to **Primitives** and drop down to **Sphere**.

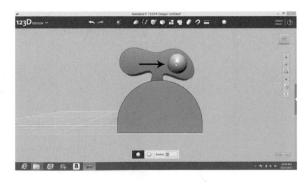

Left-click and release. Move the sphere to the desired location on the right side of the eyestalk.

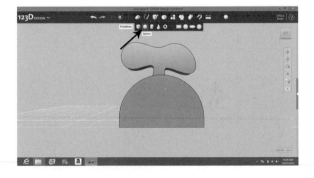

Left-click and release.

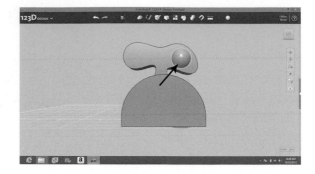

Left-click and hold.
Move the cursor down
until the sphere shrinks
slightly. This means the
sphere is being moved
into the eyestalk.
Release the left-click.

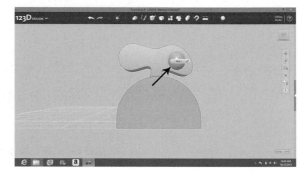

Go to **Primitives** and
drop down to **Sphere**.

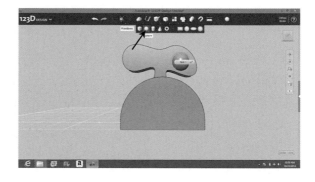

Left-click and release.
Move the sphere to the
desired location on the
left side of the eyestalk.

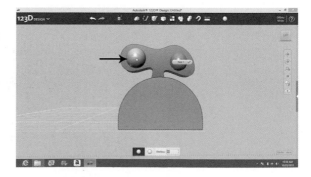

Left-click and release.

Move the cursor slightly until the sphere is outlined with a thin circle. Left-click and hold. Move the cursor down until the sphere starts to shrink (move back into the eyestalk). Release the left-click.

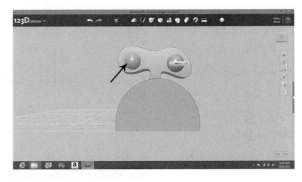

Move the cursor to the view cube and left-click on the arrow that points to the bottom. Move the cursor to the upper right corner of the view cube (left-pointing arrow).

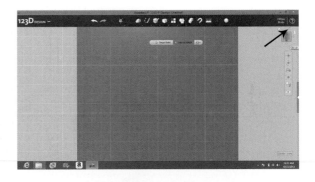

Left-click and release. Move to **Sketch** and drop down to **Rectangle**.

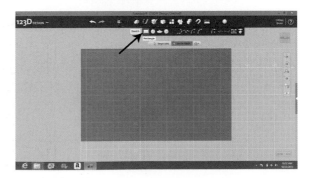

Left-click and release. Move the cursor into the rectangle.

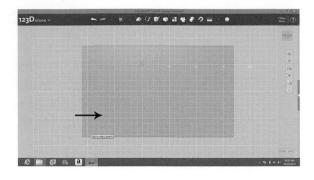

Left-click and release. Move the cursor to a point 5 mm right and 5 mm up from the lower left corner. Left-click and release. Move the cursor right and slightly up.

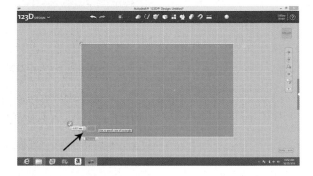

Type 126, then press **Tab**. Type 76.

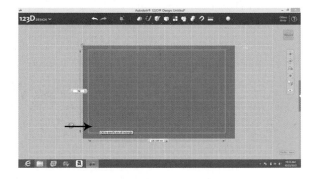

Left-click and release. Go to **Construct** and drop down to **Extrude**.

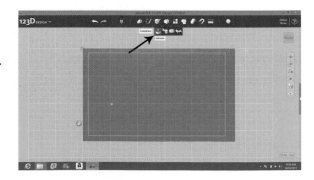

Left-click and release.
Move the cursor inside
the inner rectangle.
Left-click and release.

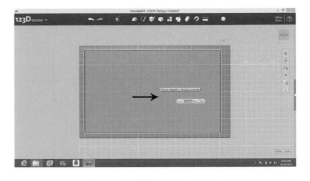

Type −5 to create an
indentation.

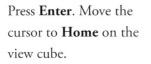

Press **Enter**. Move the
cursor to **Home** on the
view cube.

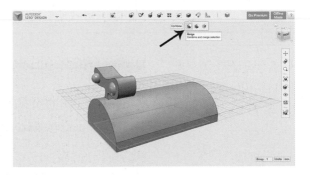

Left-click and release.
Because we added
objects—eyeball
spheres—rather than
shapes, they must be
combined to create
one piece. Move to the
**Combine** icon. Drop
down to **Merge**. Left-
click and release.

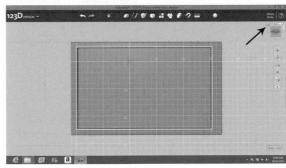

Move the cursor to the body. Left-click and release.

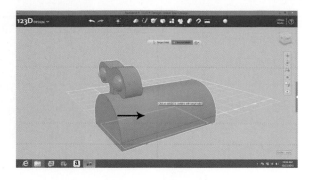

Move the cursor to the left eyeball.

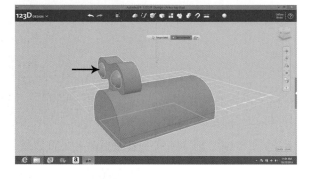

Left-click and release. Move the cursor to the right eyeball. Left-click and release. Press **Enter**. Go to **123D** and drop down to **Export STL**. Save this file to print the top: go to **123D** and drop down to **Save**.

## Clicker Base

Start a new project. Go to **Primitives** and drop down to **Box**.

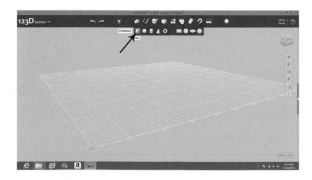

Left-click and release. Move the box to the lower left part of the screen. Type 75, then press **Tab**. Type 125, then press **Tab**. Type 70.

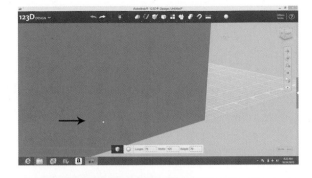

Press **Enter**. Move the cursor to **Front** on the view cube.

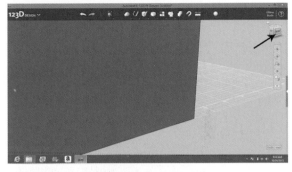

Left-click and release. Go to **Sketch** and drop down to **Circle**.

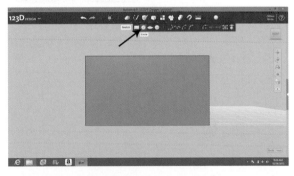

Left-click and release. Move the cursor into the rectangle. Left-click and release. Move the cursor to a point 50 mm right and 30 mm up from the lower left corner (10 grid marks right, 6 grid marks up). Left-click and release. Move the cursor slightly.

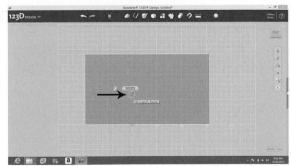

Type 16.

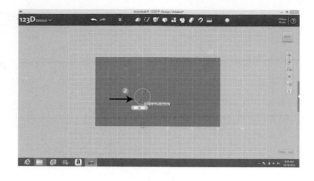

Left-click and release.
Go to **Construct** and
drop down to **Extrude**.

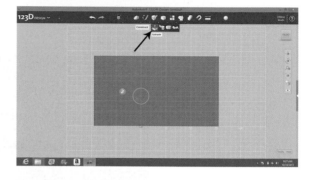

Move the cursor into
the circle. Left-click
and release.

Type −75 to create the
holes for the axles.

Press **Enter**. Move the cursor to the left arrow on the view cube.

Left-click and release. Go to **Sketch** and drop down to **Rectangle**.

Left-click and release. Move the cursor inside the rectangle. Left-click and release. Move the cursor 10 mm right and 30 mm up from the lower left corner (2 grid marks right, 6 grid marks up). Left-click and release. Move the cursor right then slightly up.

Type 55 for the width, then press **Tab**. Type 1.

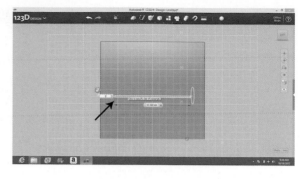

Left-click and release. Go to **Construct** and drop down to **Extrude**.

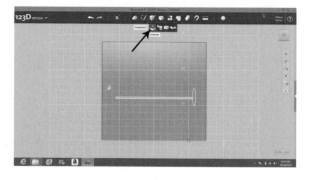

Left-click and release. Move the cursor inside the thin rectangle you just created. Left-click and release.

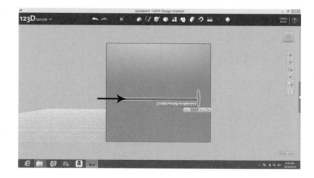

Type -10 to create the front slot.

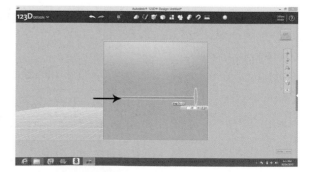

Press **Enter**. Move the cursor to the top arrow on the view cube.

Left-click and release. Move the cursor to the upper right corner of the view cube, on the left-pointing arrow.

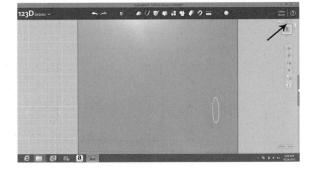

Left-click and release. Go to **Sketch** and drop down to **Rectangle**.

Left-click and release. Move the cursor inside the rectangle. Left-click and release.

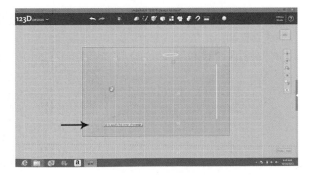

Move the cursor to a
point 5 mm right and
5 mm up from the lower
left corner (1 grid mark
right, 1 grid mark up).
This does not have to be
precise. Left-click and
release. Move the cursor
right, then slightly up.

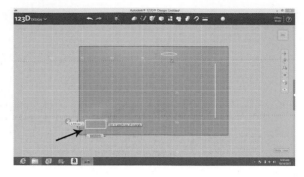

Type 115 for the
width, then press **Tab**.
Type 65.

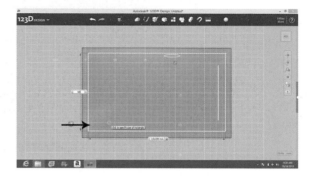

Left-click and release.
Go to **Construct** and
drop down to **Extrude**.

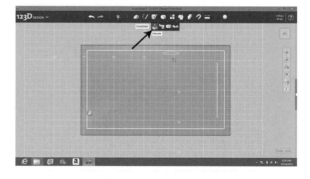

Left-click and release.
Move the cursor into
the inner rectangle.
Left-click and release.

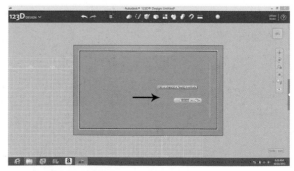

Type -65 to create the hollow interior.

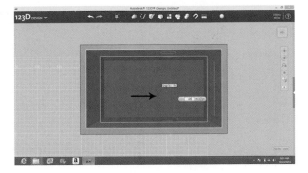

Press **Enter** and move the cursor to **123D** in the upper left corner. On the drop-down menu, select **Export STL** and save your file for printing.

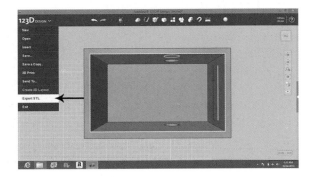

To save the design file, move the cursor to **123D** in the upper left corner and drop down to **Save**.

## Wheel

Start a new project. Go to **Primitives** and drop down to **Cylinder**.

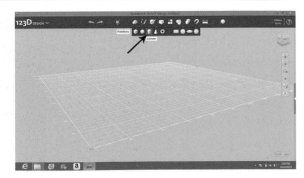

Left-click and release. Move the cylinder to the lower left portion of the screen. Type 50, then press **Tab**. Type 15.

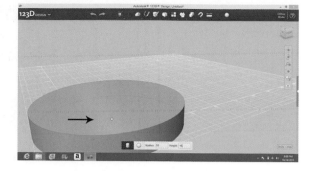

Press **Enter**. Move the cursor to **Top** on the view cube.

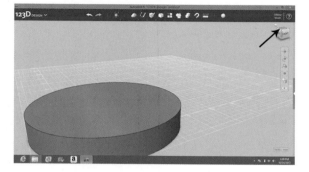

Left-click and release. Move to **Primitives** and drop down to **Cylinder**.

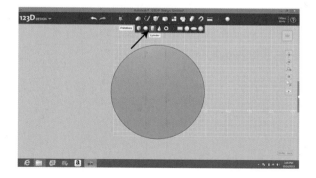

Left-click and release. Move the cylinder to the center of the circle. (It will "snap" into place.)

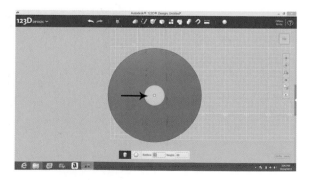

Type 12.5, then press **Tab**. Type 15.

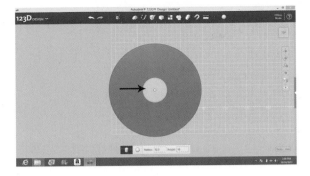

Press **Enter**. Go to **Primitives**, then drop down to **Cylinder**.

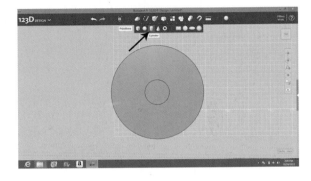

Left-click and release. Move the cylinder to the center of the circle.

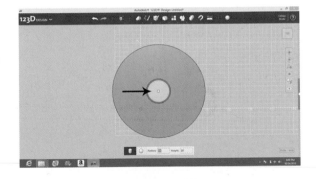

Type 7, then press **Tab**. Type 30.

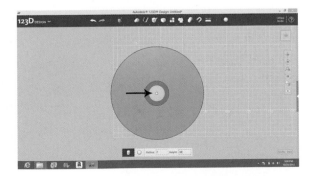

Press **Enter**. Move the cursor to **Home** on the view cube.

Left-click and release. Go to **Combine**.

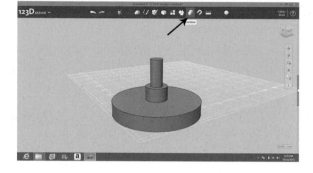

Left-click and release. Move the cursor to the top part of the axle.

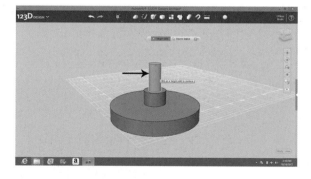

Left-click and release. Move the cursor to the middle part of the wheel assembly.

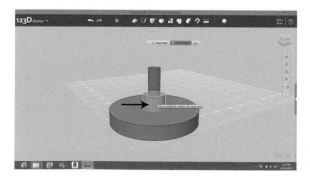

Left-click and release. Move the cursor to the bottom of the wheel assembly.

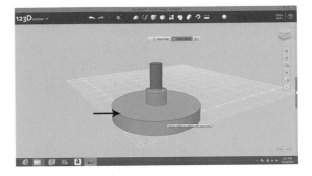

Left-click and release. Press **Enter** and move the cursor to **123D** in the upper left corner. On the drop-down menu, select **Export STL** and save your file for printing.

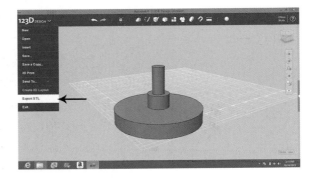

To save the design file, move the cursor to **123D** in the upper left corner and drop down to **Save**.

## Clicker

Start a new project. Go to **Primitives** and drop down to **Cylinder**.

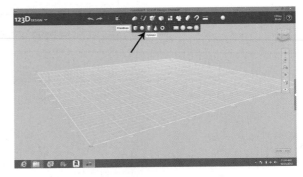

Left-click and release. Move the cylinder to the lower left part of the work area. Type 9, then press **Tab**. Type 50.

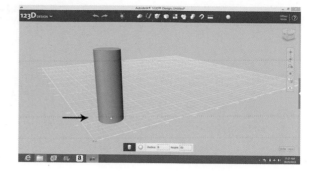

Press **Enter**. Move the cursor to **Top** on the view cube.

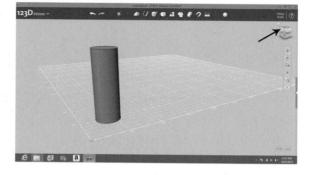

Left-click and release. Go to **Sketch** and drop down to **Circle**.

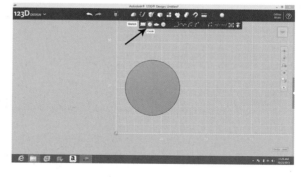

Left-click and release. Move the cursor inside the circle. Left-click and release. Move the cursor to the center of the circle. Left-click and release. Move the cursor slightly.

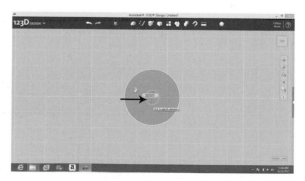

Type 14.5.

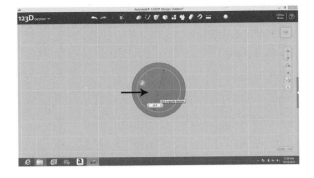

Left-click and release. Go to **Construct** and drop down to **Extrude**.

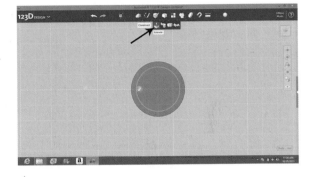

Left-click and release. Move the cursor to the inner circle. Left-click and release.

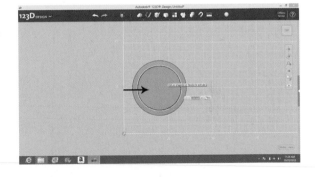

Type -50 to create the opening for the axles.

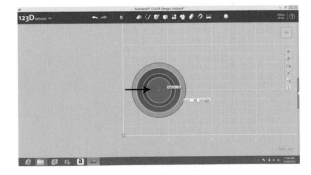

Press **Enter**. Go to
**Sketch** and drop down
to **Rectangle**.

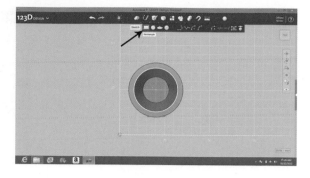

Left-click and release.
Move the cursor inside
the outer ring of the
circles.

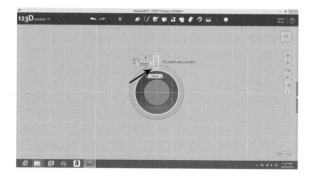

Left-click and release.
Select a place inside
the outer ring, near
the top. Left-click
and release. Move the
cursor up, then slightly
right.

Type 2, then press
**Tab**. Type 11.

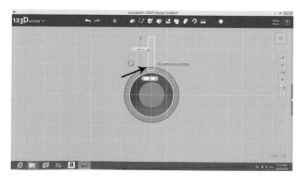

Left-click and release. Move the cursor to the "3 o'clock" position inside the outer ring. Left-click and release. Create a rectangle 2 mm by 11 mm. Left-click and release.

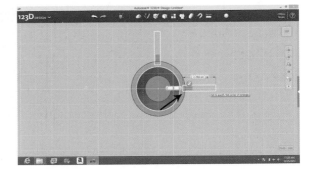

Create rectangles at the "6 o'clock" and "9 o'clock" positions.

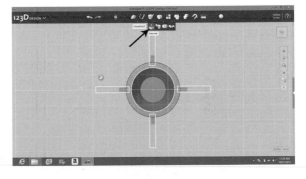

Go to **Construct** and drop down to **Extrude**.

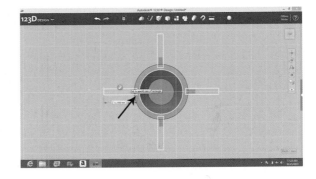

Left-click and release. Place the cursor inside one of the rectangles. Left-click and release. Move the cursor to each of the other three rectangles. Left-click and release inside each rectangle.

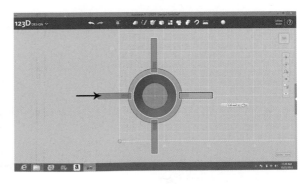

Move the cursor to
the right end of the
white number box.
Left-click and release.
Press **Backspace** until
all numbers and letters
are gone.

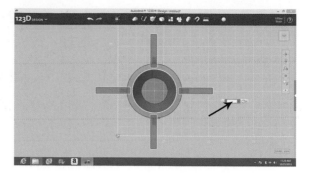

Type −50.

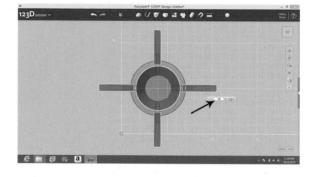

Move the cursor to the
second black arrow to
the right of the "-50."
Left-click and release.
Drop down to **Merge**.

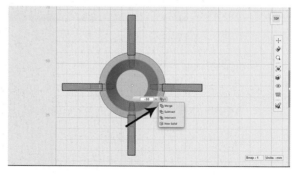

Left-click and release.

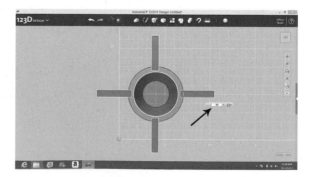

Press **Enter** and move the cursor to **123D** in the upper left corner. On the drop-down menu, select **Export STL** and save your file for printing.

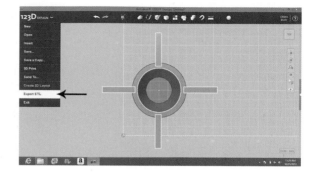

To save the design file, move the cursor to **123D** in the upper left corner and drop down to **Save**.

Insert the wheels through the holes in the side of the base. The "clicker" will hold the axles together. A business card inserted through the front slot should engage the clicker.

The top is mainly for decoration, but it provides a smooth surface for small hands.

Paint the eyeballs and add a couple of paint dots for the nose to make the toy more interesting.

# MAKING LITTLE CLICKER USING SKETCHUP

## Decorative Top

Start a new project. Go to **Camera**, then drop down to **Standard Views**. Go to **Front**, then left-click and release.

Go to the **Arcs** icon. Select **2 Point Arc** from the drop-down menu. Left-click and release. Move the pointer to the intersection of red and blue lines. Left-click and hold. Move the cursor right. (Stay on the red line.) Type 85, then press **Enter**. Move the pointer up. Type 45, then press **Enter**. Release the left-click.

Go to the **Zoom** icon. Left-click and release. Scroll until the arc fills about one-third of the screen (the lowest third).

Go to **Lines** (third icon from left), then select **Line** from the drop-down menu. Left-click and release.

Move the pointer to the left end (bottom) of the arc. Left-click and hold. Move the pointer to the right end (bottom) of the arc. Release the left-click. A filled white semicircle should appear.

Go to **Lines** and drop down to **Freehand**. Left-click and release. Move the pointer to the top edge of the arc. Left-click and hold. Draw the eyestalk—it should look similar to the one on page 258. Release the left-click when the pointer is touching the top of the arc. A filled white eyestalk should appear.

Go to the **Push/Pull** icon. Left-click and release.

Move the cursor into the lower third of the semicircle. Left-click and hold. Move the cursor upward. Type 140, then press **Enter**. Release the left-click.

Move the cursor into the upper third of the eyestalk. Left-click and hold. Move the cursor downward. Type 30, then press **Enter**. Release the left-click.

Move the cursor slightly within the eyestalk. Left-click and hold. Move the cursor downward. Type 10, then press **Enter**. Release the left-click. This offsets the front of the eyestalk from the face.

Go to the **Shape/Circle** icon. Left-click and release. Move the cursor into the right end of the eyestalk. Left-click and hold. Move the cursor. Type 5, then press **Enter**. Release the left-click.

Move the cursor into the left end of the eyestalk. Left-click and hold. Move the cursor. Type 5, then press **Enter**. Release the left-click.

Go to the **Push/Pull** icon. Left-click and release.

Move the cursor into the left circle. Left-click and hold. Move the cursor left. Type 3, then press **Enter**. Release the left-click.

Move the cursor into the right circle. Left-click and hold. Move the cursor right. Type 3, then press **Enter**. Release the left-click.

Go to **Camera**, then drop down to **Standard Views**. Go to **Bottom**, then left-click and release.

Go to **Zoom Extents** icon. Left-click and release.

Go to **Tape Measure** icon. Left-click and release. Move the pointer to the lower left corner. Left-click and hold. Move the cursor right. Type 5, then press **Enter**. Release the left-click. Move the pointer to the guide point you just created. Left-click and hold. Move the cursor up. Type 5, then press **Enter**. Release the left-click.

Go to the **Shape/Rectangle** icon. Left-click and release.

Move the pointer to the last guide point you created. Left-click and hold. Move the cursor right and slightly up. Type 76,126. Press **Enter**, then release the left-click.

Go to the **Push/Pull** icon. Left-click and release. Move the cursor into the upper third of the inner rectangle. Left-click and hold. Move the cursor down. Type 5, then press **Enter**. Release the left-click.

Move the cursor to the **Select** icon. Left-click and release. Go to **Edit**, then drop down to **Select All**. Left-click and release.

Go to the menu at the top and choose **File**, then drop down to **Export STL**. Left-click. This file can be used to print the top for Little Clicker.

To save the design file, choose **File**, then **Save**.

## Base

Start a new project. Go to **Camera**, drop down to **Standard Views**, then go to **Top**. Left-click and release.

Go to the **Shape/Rectangle** icon. Left-click and release.

Move the pointer to the intersection of red and green lines. Left-click and hold. Move the cursor right and slightly upward. Type 125, 75, then press **Enter**. Release the left-click.

Go to the **Zoom Extents** icon. Left-click and release.

Go to the **Push/Pull** icon. Left-click and release. Move the cursor into the lower third of the rectangle. Left-click and hold. Move the cursor upward. Type 70, then press **Enter**. Release the left-click.

Go to the **Tape Measure** icon. Left-click and release. Go to the lower left corner. Left-click and hold. Move the cursor to the right. Type 5, then press **Enter**. Release the left-click. From the current point, left-click and hold. Move the cursor upward. Type 5, then press **Enter**. Release the left-click.

Go to the **Shape/Rectangle** icon. Left-click and release. Move the pointer to the last guide point you made. Left-click and hold. Move the cursor right and slightly upward. Type 115, 65, then press **Enter**. Release the left-click.

Go to the **Push/Pull** icon. Left-click and release. Move the cursor into the lower third of the center rectangle. Left-click and hold. Move the cursor upward. Type 65, then press **Enter**. Release the left-click.

Go to **Camera** and drop down to **Standard Views**, then **Left**. Left-click and release.

Go to the **Zoom Extents** icon. Left-click and release.

Go to the **Tape Measure** icon. Left-click and release. Go to the lower left corner. Left-click and hold. Move the cursor to the right. Type 10, then press **Enter**. Release the left-click. From the current point, left-click and hold. Move the cursor upward. Type 30, then press **Enter**. Release the left-click.

Go to the **Shape/Rectangle** icon. Left-click and release. Move the pointer to the last guide point you made. Left-click and hold. Move the cursor right and slightly upward. Type 55, 1, then press **Enter**. Release the left-click.

Go to the **Push/Pull** icon. Left-click and release. Move the cursor until the thin rectangle is highlighted with small black dots. Left-click and hold. Move the cursor slightly upward. Type 5, then press **Enter**. Release the left-click.

Go to **Camera** and drop down to **Standard Views**, then **Front**. Left-click and release.

Go to the **Tape Measure** icon. Left-click and release. Go to the lower left corner. Left-click and hold. Move the cursor to the right. Type 50, then press **Enter**. Release the left-click. From this point, left-click and hold. Move the cursor upward. Type 30, then press **Enter**. Release the left-click.

Go to the **Shape/Circle** icon. Left-click and release. Move the pointer to the last guide point you created. Left-click and hold. Move the cursor slightly. Type 8, then press **Enter**. Release the left-click.

Go to the **Push/Pull** icon. Left-click and release. Move the cursor into the lower third of the circle. Left-click and hold. Move the cursor slightly upward. Type 5, then press **Enter**. Release the left-click.

Go to **Camera** and drop down to **Standard Views**, then **Back**. Left-click and release.

Go to the **Tape Measure** icon. Left-click and release. Go to the lower right corner. Left-click and hold. Move the cursor to the left. Type 50, then press **Enter**. Release the left-click. From the current point, left-click and hold. Move the cursor upward. Type 30, then press **Enter**. Release the left-click.

Go to the **Shape/Circle** icon. Left-click and release. Move the pointer to the last guide point you created. Left-click and hold. Move the cursor slightly. Type 8, then press **Enter**. Release the left-click.

Go to the **Push/Pull** icon. Left-click and release. Move the cursor into the circle. Left-click and hold. Move the cursor slightly upward. Type 5, then press **Enter**. Release the left-click.

Move the cursor to the **Select** icon. Left-click and release. Go to **Edit**, then drop down to **Select All**. Left-click and release.

Go to the menu at the top and choose **File**, then drop down to **Export STL**. Left-click. This file can be used to print the base for Little Clicker.

To save the design file, choose **File**, then **Save**.

## Wheel

Start a new project. Go to **Camera** and drop down to **Standard Views**, then go to **Top**. Left-click and release.

Go to the **Shape/Circle** icon. Left-click and release.

Move the pointer to the intersection of red and green lines. Left-click and hold. Move the pointer slightly, then type 50. Press **Enter**, then release the left-click.

Go to the **Zoom Extents** icon. Left-click and release.

Go to the **Push/Pull** icon. Left-click and release. Move the cursor into the lower third of the circle. Left-click and hold. Move the cursor upward. Type 15, then press **Enter**. Release the left-click.

Go to the **Shape/Circle** icon. Left-click and release.

Move the pointer to the center of the circle. Left-click and hold. Move the cursor slightly. Type 12.5, then press **Enter**. Release the left-click.

Move the pointer to the center of the circle. Left-click and hold. Move the cursor slightly. Type 7, then press **Enter**. Release the left-click.

Go to **Camera** and drop down to **Standard Views**, then go to **Iso**. Left-click and release.

Go to the **Push/Pull** icon. Left-click and release. Move the cursor to highlight the smallest circle (the one closest to the center). Left-click and hold. Move the cursor upward. Type 45, then press **Enter**. Release the left-click.

Move the cursor to highlight the next circle outward. Left-click and hold. Move the cursor upward. Type 15, then press **Enter**. Release the left-click.

Move the cursor to the **Select** icon. Left-click and release. Go to **Edit**, then drop down to **Select All**. Left-click and release.

Go to the menu at the top and choose **File**, then drop down to **Export STL**. Left-click. This file can be used to print a wheel for Little Clicker.

To save the design file, choose **File**, then **Save**.

## Clicker

Start a new project. Go to **Camera** and drop down to **Standard Views**, then **Top**. Left-click and release.

Go to the **Shape/Circle** icon. Left-click and release. Move the pointer to the intersection of red and green lines. Left-click and hold. Move the cursor slightly. Type 9, then press **Enter**. Release the left-click.

Go to the **Zoom** icon. Left-click and release. Scroll until the circle is large enough to work with.

Go to the **Shape/Circle** icon. Left-click and release. Move the pointer to the center of the circle. Left-click and hold. Move the cursor slightly. Type 7.25, then press **Enter**. Release the left-click.

Go to the **Select** icon. Left-click and release. Move the cursor inside the smaller circle. Left-click and release. Go to **Edit**, then drop down to **Delete**. Left-click and release.

Go to the **Push/Pull** icon. Left-click and release. Move the cursor inside the blue semicircle. Left-click and hold. Move the cursor toward the center of the circle. Type 50, then press **Enter**. Release the left-click.

Go to the **Shape/Rectangle** icon. Left-click and release. Move the pointer to a position inside the white semicircle, near the "12 o'clock" position. Left-click and hold. Move the cursor out from the circle and slightly right. Type 11,2, then press **Enter**. Release the left-click. Create a similar rectangle at the "3 o'clock," "6 o'clock," and "9 o'clock" positions.

Go to the **Push/Pull** icon. Left-click and release. Move the cursor inside one of the rectangles. Left-click and hold. Move the cursor slightly toward the center of the circle. Type 50, then press **Enter**. Release the left-click. Repeat this process for the other three rectangles.

Move the cursor to the **Select** icon. Left-click and release. Go to **Edit**, then drop down to **Select All**. Left-click and release.

Go to the menu at the top and choose **File**, then drop down to **Export STL**. Left-click. This file can be used to print a clicker for Little Clicker.

To save the design file, choose **File**, then **Save**.

# GOING FURTHER

## 3D Scanners

The 3D scanner, an emerging technology, is now available for consumers for less than $400. With a 3D scanner, you can make an object using clay, then scan and manipulate it for a 3D printer. For example, I made a dollhouse lamp using clay (center), then scanned and printed it (right), then reduced the lamp to doll size (left).

Complex items (if the details are not too small) can be scanned and combined with software creations. For example, I scanned a seashell, then combined it with a ring (similar to the one designed in this book). The STL files for the scanned shell and the ring can be combined in TinkerCad (free software). Simply upload the shell and ring files (.stl), then move them around until you like the combination. Merge the files, export, and you'll be ready to print.

You can even scan your own head and merge it to a doll's body. Imagine making dolls that are replicas of family members!

## Now You're on Your Own

Now that you are ready to design and make your own things, I would suggest you start by making a sketch of the front, side, and top views of your object. Mark dimensions on the sketch so you will know how much to cut or extrude. A caliper (which costs less than $15) is very useful for measuring.

Think about the way things will fit together. If you want to make a hole for a bolt, make the hole slightly larger than the bolt. There will be some trial and experimentation here, depending on the accuracy of your printer and how snug a fit you desire.

Print out your design. When you have it in your hand, it's easy to see the modifications that need to be made for improvement. Go back to your design file, make changes, then print again.

You are now part of the next industrial revolution in manufacturing.

# Also by Mike Rigsby

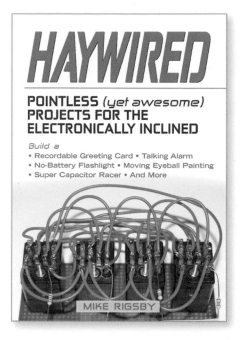

## Haywired

Pointless (Yet Awesome) Projects
for the Electronically Inclined

978-1-55652-779-1
$16.95 (CAN $18.95)

"A great first book for the
budding inventor or engineer."

—Eric Wilhelm, cofounding
partner of Squid Labs,
CEO of Instructables

Written for budding electronics hobbyists, *Haywired* proves that science can inspire odd contraptions. Create a Mona Lisa that smiles even wider when you approach it. Learn how to build and record a talking alarm or craft your own talking greeting card. Construct a no-battery electric car toy that uses a super capacitor, or a flashlight that can be charged in minutes then shine for 24 hours. Each project is described in step-by-step detail with photographs and circuit diagrams, and helpful hints are provided on soldering, wire wrapping, and multimeter use.